A DOCTOR'S JOURNEY

A Hungarian's Realization of the American Dream

Laszlo Makk, MD, FCAP

Order this book online at www.trafford.com
or email orders@trafford.com

Most Trafford titles are also available at major online book retailers.

Printed in the United States of America.

ISBN: 978-1-4269-3801-6 (sc)
ISBN: 978-1-4269-3802-3 (hc)
ISBN: 978-1-4269-3803-0 (e)

Library of Congress Control Number: 2010913053

Trafford rev. 11/13/2010

www.trafford.com

North America & international
toll-free: 1 888 232 4444 (USA & Canada)
phone: 250 383 6864 • fax: 812 355 4082

This memoir is dedicated to the memory of
my beloved wife, Carolyn,
and my wonderful family.

ACKNOWLEDGMENTS

I am most grateful to my parents, Istvan and Ilona Makk, who set a fine example for me with their hard work, integrity, pursuit of knowledge, and love. I have tried to follow their example throughout my life.

Very special thanks to John and Faye Bender for taking me into their home when I was a penniless and homeless immigrant and for treating me and my family as part of their family.

Deepest gratitude goes to my family and friends for their love and encouragement, and for urging me to write my life story, especially to Lloyd Hilton Smith, who first suggested that I publish my story back in 1967. Special thanks to Albany Medical College for letting me continue my medical studies and to Texas Medical Center for the outstanding medical specialty training I received there.

For the special assistance they provided with this publication, I want to thank my daughter-in-laws Carolyn Mock Makk and Catherine Makk, as well as Rene von Richthofen and his lovely wife, Jane Manus. Thanks also to Sasha Lawrence, who transcribed my original, barely legible, handwritten manuscript.

Contents

PROLOGUE

It was a hot summer Sunday afternoon in the Hungarian town of Pannonhalma. It was 1938, and I was only six years old. I was "helping" my father fix our gate post when the doctor came down the hill toward us. He seemed disturbed and was heavily perspiring. My father helped him sit down in the shade and offered him a glass of cold water. He then told us what had happened.

He had just come from treating some stab wounds among indigent family members who often got drunk and fought with one another. Because they had no money, the doctor would often provide help for free. This had not been the first time the doctor had been called to help out this family, so he was frustrated. The doctor told my father he could not even have his Sunday rest because of some drunks fighting.

It was then that the doctor turned to me and asked, "What are you going to be when you grow up?"

I told him, "I want to be a doctor."

I don't know what made me say that, but from that day forward, I wanted nothing else but to be a great doctor some day.

From then on, anything medical interested me. I took a piece of board and put a nail in each corner. I caught a frog, tied its legs to the nails, and opened it up to see how its heart beat. Every time something happened in the health area, I had to see it. When two men died while they were building the gymnasium adjacent to the Pannonhalma Archabbey, I felt compelled to see them. Their bodies had been placed in the cemetery's chapel, but the doors had been locked. So I stood on my friend Feri's shoulders to peep through the window just to find their remains covered by sheets.

At times, it was not easy to transform this commitment into reality. There were many obstacles along the way, but I never gave up, however impossible it became to get through them. Hard work, determination, hope, and good luck kept me going.

Chapter I

The Beginning

Isten áld meg a magyart.
"God bless the Hungarians"*

Hungary is a small country in Central Europe. In western Hungary, there is a Benedictine Archabbey named Pannonhalma. The first king of Hungary, St. Stephen, founded it in 996. Before King Stephen's reign, the Hungarians were nomads, pagans, and excellent horsemen. Their livelihood depended on raiding and robbing their prosperous western neighbors. During these raids, the best and bravest were usually killed, and King Stephen was concerned that the lame and hunchbacks, who were incapable of fighting, would consequently amass the future generations of his people. King Stephen noted the West European countries had wealth, agriculture, trades, and Christianity.

He converted his nation to Christianity in order to help his people settle down, learn agriculture and trades, and create a stable nation. He sent an emissary to Rome and requested a crown from the Pope so he could be king. He also requested some missionaries to convert the Hungarians to Christianity and to teach them various trades and agriculture.

The Pope sent Benedictine monks in 996, and King Stephen settled them in western Hungary on a hilltop called Pannonhalma located 65 miles west of Budapest. They built a monastery and established an archabbey. King Stephen gave the monks large land holdings and power. Eventually, a town was established in the foothills of Pannonhalma. Many of the

* *the first line of the Hungarian National Anthem*

townspeople made their living by working for the monastery or on its extensive land holdings. The town was called Győrszentmárton, deriving its name from Saint Martin of Tours. According to legend, it was here that St. Martin, a Roman soldier at the time, gave half of his robe to a beggar.

My grandfather, Pàl Makk, was a stonemason like many Makks before him. He stepped on a rusty nail and suffered a horrible death from tetanus. He left a widow and four young children, one of which was Istvan Makk. The young widow decided not to let her children be taken away to an orphanage and kept the family together. They all had to go to work. Istvan had to drop out of the fourth-grade grammar school and go to work pulling weeds in the fields at age ten. In those days, the foreman was on horseback behind them with a long whip. The foreman whipped the one who was last in the weed-pulling row. Istvan learned from this that if he wanted to avoid getting whipped, he had to pull the weeds faster and work harder to stay near the front of the rows. Consequently, a hard work ethic stayed with him throughout his life.

A few years later, Istvan became a shoe and boot maker's apprentice. By the time he was twenty, he had become a master boot maker and had opened his own shop. He married Ilona Simon when she was 22 years old. A well-educated farmer's daughter, Ilona was a beauty from the next village. She walked two kilometers daily to and from the railroad station to take a two-hour roundtrip train ride to the nearby city to get a good education from a higher school. She was an avid reader, and as a deeply religious person, she raised us to be devout Catholics. This was the background of the Makk family when little Laszlo was born on April 16, 1932.

It had been a nice spring day in Győrszentmárton, but it was not over yet. As the sun was setting and the noises of work slowly died down, the kerosene lamps were lit in the small houses to provide light for the evening meal.

The bell ringer for the church checked his pocket watch and briskly walked toward the bell tower to toll the bells at exactly eight o'clock to remind the villagers that it was time for the evening prayers and then bed. After the bells stopped ringing, the kerosene lamps were blown out one by one, and people turned in for the night. But this was not the case for the Makk family. Their kerosene lamps stayed on, and there was feverish activity in the house. Ilona Makk was about to give birth to her fifth child. Everyone prayed that the child would live. Of the Makks' previous four children, only one had survived. Finally, the child was born alive, and it was a boy. The baby was in this world for less than half an hour when a large green spider started to descend from the ceiling toward him. The parents, the midwife, and the others in the room all thanked God for a living son. They all felt the green spider was a sign of good luck and prayed and felt that great things would happen to the newborn son. That child was me. I was named Laszlo after King St. Laszlo, who chased the Tartars out of Hungary and saved Christianity for Europe. For this brave act, King Laszlo was canonized.

My childhood was quite exciting. My parents built a new two-story home, which unlike others was designed by an architect. To save money, my father would haul up the bricks, water, and mortar onto the scaffolding after work and at night so the bricklayers could start building first thing in the morning and not spend time gathering materials. My parents also raised pigs to help pay for the house. Everything my father did was high quality. His pigs were of such a high quality that a butcher from the city would purchase them and pick them up with a truck. This was a big sensation, because at that time, the only other truck that came to town was the Shell kerosene truck. The next big excitement for my family was moving into our new house, which had electricity and room for everything. When we moved in, the first thing my mother brought was a kitchen stool with a crucifix and bread and salt on it. That same crucifix hangs today in our home in Palm Beach, Florida.

The happiness in the new Makk home didn't last long. After Hitler's Nazi regime annexed Austria and occupied Czechoslovakia, Hungary was next on his agenda. It was just a matter of time before the Nazi monster would occupy little Hungary, who bravely decided to resist the Nazi steamroller and wanted to stay neutral. Our hometown was in the contemplated area of the frontier. One day, the town put out a declaration that everybody had to be prepared to evacuate with a 15-minute notice. The man of the house could no longer be expected to accompany his family, because he would be drafted into the defense force. No transportation aids would be available, either. The only thing evacuees could take was what they could carry themselves. I was supposed to carry a bag of sugar and a jug of water. My brother, Tibor, would carry the blankets, and my mother would carry my two little sisters, Zsuzsanna and Veronica. Of course, I couldn't understand why we would have to leave our beautiful new home. With tears in her eyes, my mother kept saying that it was all over, that my father would likely die defending our country, that our family would be homeless refugees on the run and most likely separated from each other.

We didn't need monster movies. We had the real monster, Nazi Germany, ready to crush us.

The war with Germany never came. When Hitler found out about little Hungary's plan to resist, he invited the Regent of Hungary Miklós Horthy for a state visit. I still remember listening to the radio and hearing a band play as Horthy's train arrived in Berlin. Hitler played the gracious host. The Hungarian head of state explained that Hungary wanted to remain neutral and wanted no foreign troops of any kind on its soil. Hitler assured him that he understood. As he was leaving, the military band played again.

Shortly after Horthy's train left Berlin, it was switched to an abandoned railroad track. At the same time, German troops started to pour into Hungary, falsely claiming that they were coming based on a treaty with the *Führer* and Horthy. There was great confusion, and sporadic resistance was annihilated. The prime minister knew the truth about the German lies, but he was unable to take any action before the Germans had shot and killed him in his office. It was falsely reported that he had committed suicide. They put a Nazi puppet in his place. It took less than a day for the German Army to take over Hungary.

In my hometown, we did not know anything about this. The first thing we did notice was German motorized army units taking over the town. I also noticed that they had excellent motor vehicles. Some of the soldiers had black uniforms with "SS" on the collars. They were part of the infamous and greatly feared Schutzstaffel organization. I also noticed that when a Hungarian colonel came to inquire, the German soldiers did not salute him but ignored him. Military etiquette and civility was gone.

Chapter II

Schools and World War II

I recall many things about my childhood. For example, while I was growing up, I was a skinny kid. My parents were very concerned, and our doctor told them I probably had tuberculosis and needed to eat a lot of spinach. I hated spinach. It made me gag, and I couldn't swallow it. Sometimes my mother stood behind me with a wooden spoon and hit me if I didn't eat the spinach soufflé that she had carefully prepared. Then the doctor told them I should drink Ovaltine, but I couldn't swallow that either. Eventually, I grew up and was healthy without needing any of these things.

I remember when it was time for me to go to kindergarten. The teacher was a strict nun. A strong disciplinarian, she walked around with a ruler and hit us in the head or fingers if we didn't do something right. Once I was an angel in a Christmas show we were having. During the show, we had to stay put and wait until the priest arrived; however, he was very late, and I needed to go to the bathroom badly. I asked the sister again and again to let me go, but she told me to hold it. Just as the priest arrived, I started to wet my angel costume and the floor. Sister told me that I had ruined the Christmas show, and she was never nice to me after that.

One day in the spring, the children at school played wolf and sheep. We were supposed to form a circle, but the circle was crooked, so the nun kept running around making us step forward and backward. As she was running around, the sister stepped on her habit, fell, and rolled over. We could see her long underwear. Nobody dared to move, but my friend Feri and I broke out into laughter. That was it. She dismissed the class—except for the two of us. As a penalty, she locked us in the classroom and told us we would be there all night when the witches and vampires would come to get us. We were scared to death and cried for a long time. After a while, we realized that if we put a chair on our little table and pushed it to the window, we could reach the latch, open it, and escape. Finally, we opened the window, and on the way out, we had to crawl through their tulip garden while the Sisters were eating lunch. We broke every tulip we could reach and ran home. I was scared all day of what would happen the next morning when I would have to go back to kindergarten, as I knew I would be severely punished.

During breakfast, I slipped out of the kitchen and hid behind the woodpiles in the basement. My mother kept looking and calling for me, but I stayed quiet. Then my mother sent for my father, because she thought the gypsies might have stolen me. Finally, I came out of hiding, crying, and told my parents what had happened. I also told them that I'd rather die than go back to kindergarten. My father said, "If you don't go, I will take you. You have to be responsible for what you have done."

My mother was more sympathetic and told my father, "He'll start school in September and will be in school for a long time. Why don't we just let him quit and enjoy the time off?"

My father relented, and I thought my mother had saved my life. It was a great spring and summer, not having to go to kindergarten. By the fall, I was ready to go to school, which I liked very much.

In Hungary, grammar school went up to the sixth grade. High school, known as gymnasium, lasted eight years. Most people in those days didn't go to gymnasium. A child could start gymnasium after he or she completed the fourth, fifth, or sixth grade of grammar school and passed an exam. I qualified after fourth grade. My parents and I desired that I attend the famous Benedictine Gymnasium at the top of the hill attached to the Archabbey of Pannonhalma, but it was not in the cards. My admission was denied even though I qualified. The principal told my father, "Mr. Makk, you already have a son in high school, and if the second boy is going to go to high school, who is going to make our shoes? Make him a shoemaker." I could tell my father was very sad about this response.

My father, who had never had a chance for education, wanted badly to educate us so that we would become "learned men" and wouldn't have to work as hard as he had to in order to make a living. The next high school was about fifteen miles away in the city of Györ. I was readily admitted there; however, to go there, I had to get up every morning at 5:45 to walk to the railroad station, which was about one kilometer away, catch the 6:30 train, and arrive in Györ by 7:30 a.m. After school, I caught the two o'clock train back and was home by 3:30 p.m. to get started on my homework. This was not easy in rain or snow, and I was just 10 years old at the time.

Hungary then entered World War II, and during school hours, we had to practice air raid evacuations to shelters. Located next to neighborhood houses, these shelters were ditches in the ground with logs and dirt on top. Because they were small, they could only accommodate a few people. We were assigned to different shelters and had to practice getting to them in a hurry.

As the war progressed, the air raids increased, because Györ was an important railroad center. Our gymnasium was taken over and converted into a military hospital, and we had to attend school in the evening from two to six o'clock at another school. This meant that I wouldn't get home until around 8 o'clock at night.

It was a heartbreaking experience to hear the screams of the wounded and dying soldiers as we walked by our old school. Our gym was even turned into a large morgue. This was not a conducive atmosphere for learning. Our train was often sidetracked and delayed because some of the rails had been bombed and damaged, and the German military trains had priority. This was an important route for the Germans to deliver supplies to the Russian front. Every time the Allies bombed the railroad, it was rebuilt quickly, because the German army and Hungarian railroad workers would work around the clock to make the repairs. When it was finished, the Allies just bombed it again. The Allies must have had good spies on the ground who would notify them when the railroad had been rebuilt.

The Allied air forces came to bomb the railroads in Györ on weekdays and hit their targets very well. The Soviets flew at night, and they had large flares. We called them Stalin's Christmas trees. They dropped their bombs randomly, more on civilian populations. After a while, we learned to go to air raid shelters at night and to stay home during the daytime raids. The Allies used the monastery at the top of the hill as their turning point for the bombing run to Györ. B-17s and B-24s flew in groups of 27 from Italy. After a while, we noted that smaller fighter planes accompanied the big bombers. My father and his friends figured out that the Allies must have been advancing closer, because the small fighters had shorter flying ranges. Sometimes we saw spectacular fights between the Allied and German Air Forces.

My father's back eventually became very troublesome, and he could barely walk. The doctors couldn't figure out what was wrong with him. My brother, who was seventeen then, was drafted into the army. Consequently, at age 12, I became the breadwinner for the family. I went to the farmers and traded bacon, lard, hams, flour, and schnapps for bed sheets, tablecloths, and wool. I traded the schnapps with the German guards for wool from their storage and took it back to the farmers for trade.

One morning in the summer of 1944, my father came home, closed every window, and turned on the radio. The newscaster on Radio Bari, an Allied station in Italy, told us that the invasion of Western Europe had begun that morning in France. My father said that the Germans wouldn't last much longer and that we should not show any signs of joy, because it could cost us dearly, especially if seen by the ever-watching Nazis.

During the war, German soldiers would sometimes take over part of our house. The last German soldier was an SS courier who had a 500cc BMW motorcycle. On Hitler's birthday, he asked if he could listen to the *Fuhrer's* speech on the radio. We couldn't say no. As Hitler came on with his usual paranoid demagoguery, tears started to flow from the soldier's eyes. Then he showed us a picture of his five-year-old son, whom he'd never seen because he had been on the front for five years without a furlough. He said his only hope for living was to be captured by the Allies on the western front. If the Soviets captured him, he would be tortured and killed. We were afraid to react to this, because we thought he might have been setting us up and we didn't want to suffer the consequences. We pretended we didn't understand.

The year 1945 greeted us with new hope for the war's end. The Benedictine Archabbey was placed under International Red Cross protection because of its historic significance. It was in the winter of 1945 that my brother was drafted into the army. My father gave him a long-haired brush and told him to brush his right knee secretly every day for about twenty minutes. He told him that the knee would turn red and swollen and would look like bad arthritis after five or six days, and hopefully, he would then be dismissed from the army.

Just as he was ready to leave, the German army surrounded and sealed off our town and searched every house and yard, looking for somebody, but no one knew whom. Later on, we found out they had been looking for a dwarf who was a spy, but the dwarf had escaped by fastening himself under the bottom of a circus wagon. The Germans searched the wagon but failed to look underneath. We knew what was going on and hoped that the dwarf escaped, which he did as the circus moved on.

As a teenager, I had many exciting experiences of the most undesirable kind. As we heard the Soviet artillery become louder and louder, the Germans started to get desperate, and so did the Hungarian Nazis.

By this time, my father couldn't walk because of his back pain. Everybody was ordered to go dig trenches. Our doctor gave us a certificate stating that he was incapable of doing that. I took the certificate to the town's Nazi chief, who had replaced our mayor. There was an armed guard in his office, and as I presented the certificate, he started to scream at me, pushed me in a corner, and yelled that they didn't accept illness as an excuse, that someone was either dead or alive. If my father didn't show up to dig trenches, he would shot, because if my father couldn't help, he couldn't use up their food and oxygen.

As the German-Russian frontier got closer and closer, he never had a chance to send a guard to kill my father. I was nearly thirteen during the last month of the war when an announcement came that everyone my age had to report for army duty. This duty would consist of giving us fake wood rifles to man the front line. They hoped the Soviet troops would be slowed down this way, because they would have to shoot us first. Thank goodness the monks in the monastery, which was under Red Cross protection, took me in. They also gave a room to my family and gave a small room to my father to operate a shoe repair shop. They also hid a lot of Jews from the Nazis there.

During Nazi occupation, the Jewish people were ordered to wear a yellow Star of David. Our family had many Jewish friends, and we tried to help them whenever we could. Pretty soon, they were removed from their homes and relocated to a ghetto. Some of them had asked my father if he could hide some of their valuables until they had returned. This was risky, because if the Nazis had found out, we could have been arrested. But my father felt that it was our duty to help them however we could. The big question was where to hide the valuables. One of the chimneys in our house was never used. There was a door attached to it in the basement and one in the attic. Using the attic door, we filled up this chimney with our valuables and those of our Jewish friends. We put food and other valuables in the overhang of the roof. The space was very narrow, so I had to climb in there and hide the items myself.

We could also hide things at my uncle's house, which was built on the side of a hill. In the kitchen, which was against the hill itself, there was a small storage cabinet insert. Once when they were putting things in this cabinet, they discovered the wall was hollow in the back. They broke it open and found a small cellar behind the wall, I was the only one small enough to fit through that hole. In the evenings after dark, I went to our uncle's house with some of our valuables. My uncle hoisted me up, and I climbed in. I was scared to death because I only had a candle for light, and I thought that snakes or rats might have been in there. This process of hiding valuables went on night after night. Each time I came out, they covered up the hole and closed the cabinet door. They told me that if somebody were to come to the house, they would put a fake wall on the back of the cabinet, and I would have to remain absolutely quiet. When the air cleared, they would let me out. My father and uncle told me that I should never tell anyone about this, particularly the police interrogators because if they found out, we would be arrested and taken away for good. I was very proud of myself for being a part of this, and every time I saw a Nazi soldier or police officer, I thought we were smarter than they were.

Mr. Brenwaszter was a Jewish leather merchant and a good friend of my father. One of the things I remember about him was his willingness to work with my father to provide local orphans with free shoes once a year. He supplied the leather components while my father donated the labor to make them. Along with other Jewish people, Mr. Brenwaszter was taken to a concentration camp in the nearby city of Györ. One day, he somehow arranged a pass and showed up at our house with an armed German soldier. He was very hungry, and my mother fixed him a great meal quickly. His German guard was very stiff at first. He wouldn't sit down, and he would not take a drink or food. My father kept insisting that they drink to his health, and finally, he gave in and had a couple shots, put his rifle down, sat at the table with Mr. Brenwaszter, and shared the meal with him. When they left—practically friends now—Mr. Brenwazter had a look of finality on his face. We all knew that this would possibly be the last time we would see him. With tears in his eyes, he thanked my father for the meal and the courage to receive him and for taming the German guard, because he had hoped that the guard would somehow learn to treat him better. Shortly thereafter, the Nazi mayor of our town let us know that he was aware that we had welcomed a Jew in our home and that we would be watched.

Shortly after that visit, my father brought home a duckling. It was greatly welcomed, because at that stage of World War II, food was scarce. I was excited that we had a duck to eat. My parents explained that we would not eat it. Instead we would get it to the suffering Jewish people in the ghetto in Györ. My father packed the cooled duckling in a small box, and I left for Györ to go to school and deliver the duck to the ghetto. As I neared the ghetto wall at a certain street, I waited for the Nazi guard to walk in the other direction and then threw the box over the wall where Mr. Brenwaszter and his friends were waiting on the other side. If I had been caught, and even if I was tortured, I was not to tell the police anything except that a man who had now disappeared had given me the box and said that if I threw the box over the wall, he would give me a tip. With my heart beating in my throat, I used all of my strength to make sure the box cleared the wall. It did, and the guards didn't see me. Afterward, as instructed, I walked away with steady steps. I was in school by 2 o'clock, but I was far too shaken up to learn anything. My parents were anxiously awaiting me to come home after school. What an experience for a 12-year-old. I was proud of myself.

We never heard from Mr. Brenwaszter again. Maybe my smuggled duckling was his last good meal before he perished in the gas chambers. Even today, thinking about that risky act gives me great pride.

Mr. Braunschweiger was a Swiss industrialist who had volunteered his services to the International Red Cross during World War II, because Pannonhalma was under Red Cross protection. He was the Red Cross representative during the war to protect historic treasures. One day, as the German-Russian front was getting closer, a German artillery unit tried to establish a position in the International Red Cross protected neutral zone next to the walls of the monastery. We watched with great anxiety as Mr. Braunschweiger bravely faced the German commanding officer and ordered them to leave. Their discussion was heated, and we thought that Mr. Braunschweiger was going to be shot by the German officer at any moment; however, the officer didn't reach for his gun and eventually ordered his unit to move on.

That night, a fierce battle was fought between the advancing Soviets and the retreating German forces in the valley below the monastery. The next day, I peeked out through a window, trying to see if there had been any explosions near our house, because the cannon shells were getting closer.

As the Soviet front was getting closer, my parents moved into our room in the monastery, leaving nearly all of our belongings behind. The SS soldier, however, remained in our house. At that time, his food ration for a week was a can of ground-up pig and a loaf of bread with sawdust in it. Our pantry was full of hams, sausages, preserved fruits, and other foods. When the German soldier left the night before the Soviets occupied our town, he told our neighbor that he didn't lock up our house because the Soviets would most likely break down the door anyway in search of food and valuables.

Before the Soviets occupied our town, I was curious to see what things looked like, so I snuck out of the monastery to see if our house had been destroyed as the front advanced. To my great surprise, nothing had been touched in our pantry, and the starving German soldier hadn't helped himself to anything at all.

That night, there was a tremendous firefight in the valley between the Germans and Soviets. Their new guns were rocket launchers. The launcher sounded like a hurt dog when it was fired. There was a color to these rockets, and in the dark, we could see where they were firing. The next morning, despite instructions that everybody was to stay in their rooms, I snuck up to one of the classrooms in the gymnasium, where I could observe the front. As the artillery shells got closer, I decided to run back to our room. Just as I started running down the staircase, there was a huge explosion. A Soviet artillery shell hit the classroom next to me. Later, I wondered if they had spotted me and had decided to fire at me.

Chapter III

Teen Years, Soviet Occupation, and Imposition of Communist Dictatorship

Then it was relatively quiet, and Soviet soldiers appeared from everywhere. In front of the monastery, there was a group of seven or eight older German soldiers who were surrendering. The Soviets brutally beat them unconscious. When some of them started moving again, they went back and started to savagely beat their heads with their gun butts. The first Soviets we saw were dirty and unshaven, and their belongings consisted of a gun and a small sack that was tied to their back with strings. Some of them insisted upon entering the monastery despite the Red Cross protection. The monks had Russian interpreters at hand.

The Soviets demanded coffee from the welcoming committee, which included a count, Mr. Braunschweiger, and the translators. When the coffee arrived, they ordered the count at gunpoint to drink some, even though it was scalding hot. Then they waited about fifteen minutes and started to drink the coffee themselves. Apparently, this was done to assure the coffee was not poisoned.

It was fascinating to see the various Soviet fighting machines. The troops were usually filthy, uneducated, and poorly outfitted. Their trucks were either old, rickety machines on their last legs or newer ones with the letters GMC on their hoods. I did not know then, but I know now that GMC stood for General Motors Corporation. These GMC vehicles were part of a Lend-Lease program run by the United States that provided hundreds of thousands of U.S. made trucks to the Allies in Europe including the Soviet Union.

In the town, the Soviet troops raped, murdered, and robbed innocent civilians. Women dressed like old ladies or stayed at home. People started to wear their wristwatches on their ankles,. Otherwise the Soviet soldiers would have jerked the watches off their wrists.

In our gymnasium, there was a first aid station that was converted into a small hospital and staffed by volunteers, including a famous surgeon from Budapest. After the German-Russian front moved, a large number of wounded civilians were taken there for treatment. It was off-limits to us, but I managed to get inside. I loved everything about it—the smell, the white coats, the uniforms, the doctors and the nurses hard at work. I will never forget seeing a boy my age whose right leg had been blown off by a landmine. He didn't make it like so many others.

Remains of the dead were kept in the makeshift morgue set up near the monastery in an historic building (erected to celebrate the 1000-year anniversary of Hungary) until family or friends came to claim them. They would take them away in buggies or wheelbarrows. At times, the chief porter was busy and gave me the keys to open up the morgue and release the bodies, because I was not afraid of the dead people. I wondered on occasion if more could have been saved. This made me even more determined to become a doctor and save lives.

About two months after the war was over, a convoy of Soviet trucks appeared at the gate of the monastery on a Sunday morning, and they ordered the porter at gunpoint to open it. Once they were inside, they went everywhere to rob and pilfer. They got the archabbot and Mr. Braunschweiger, the Swiss representative of the International Red Cross, against the wall, their guns pointed at them, and we didn't know if they would be killed or not. We kids tried to hide chalices and other holy treasures. They found Mr. Braunschweiger's Ford automobile. A drunken Soviet got into the car, put it into reverse, and hit the wall of the monastery full blast, totaling the car.

The Soviets were most interested in alcohol, which they found plenty of and drank. We tried to copy the Soviet trucks ID numbers, but we had trouble with their Cyrillic lettering. We thought that there would be an investigation by the Red Cross for this heinous act and that it would be helpful to identify the perpetrators. We watched with broken hearts as the Soviet Army trucks (loaded up with stolen centuries-old treasures and all the wine they could find) left the monastery. They also took Mr. Braunschweiger with them.

About a month later, when we gathered for evening prayers in the chapel, one of the monks sadly informed us that the Soviet Army had executed him.

Later, a Soviet Army detachment came with a general in charge. They were going to "investigate the alleged robbery." I was one of the people questioned. Through an interpreter, I described the markings in order to identify the trucks. His final conclusion was that the perpetrators were not Soviet troops but Nazi soldiers who had been left behind. The Soviet general said the Germans were wearing Soviet uniforms, armed with Soviet guns and speaking Russian to fool people. That was the end of the inquiry. This was an absurd, impossible, incredible lie that no one believed. Then he warned us that if we continued to insist that they were Soviet troops (not Germans or Hungarian reactionaries), defaming the liberating Soviet Army, then they would see to it that we would receive the appropriate punishment.

At age thirteen, it was my first experience with Soviet-style justice. But this experience was only the beginning. The Soviet troops were often drunk and frequently raped women, sometimes in front of their children, sometimes several soldiers at a time. If a husband objected, he was shot. The Soviet troops could take anything at gunpoint.

Sandor Titrick was an orphan, and my parents took him in as an apprentice shoemaker and an adopted family member. After he got married and moved to a neighboring village, he would come on occasions and get me to stay with them. He was a fantastic hunter and sharpshooter. During Soviet occupation, nobody was allowed to keep guns or hold hunting licenses except the Communists. Sandor had a number of guns and some ammunition in the thatched roof of his home. Should the Communists have found any of them, the punishment would have been decades in a gulag-type hard labor camp. We used to go poaching together, and it was very exciting. One time, when Sandor fired at a pheasant, a lot of submachine gun fire came over our heads. Either the Communist police or Soviet soldiers hunted in the same area. Sandor told me to follow him and to hang on to my gun. We ran very hard, and when we reached a winding creek, we jumped into it and kept running. Finally, Sandor thought we were at a safe distance to come out of the creek. He told me that the reason we had to run in the creek was to lose our scent so the attack dogs couldn't follow us. Then we disassembled our guns and hid them under our coat and went back to his house in a large circle. He told me that if they tried to capture us, he wouldn't be taken alive, and he would take some Soviets with him.

These were weird experiences for a teenager under Soviet rule. On Sundays, my father sometimes borrowed a horse and buggy, and we would go to nearby villages to take orders and measurements for boots and shoes. Once we went to a village that had a lot of vineyards. Near the village were a lot of Soviet troops. We thought they were on maneuvers, but something was strange. As soon as we came upon the first farmer, he told us to unhitch the horse, put the harness in the barn, hide in the upstairs hayloft, and stay quiet and motionless until we heard from him again. Later on, he told us that two days before, two drunken Soviet soldiers came to his wine cellar and demanded wine at gunpoint. They then demanded that the farmer produce his wife. The farmer was slow to respond so the drunken soldiers started to shoot up the wine barrels. Most of the wine flowed out of the bullet holes. The farmer could not take it anymore and killed both Soviet soldiers before they killed him.

As it turned out, the Soviet troops we had seen were coming to seal off the village. They told the mayor if the farmer who had killed the Soviet soldiers didn't step forward, they would line up every man in the village and start randomly shooting them until the farmer who had killed the Soviets was produced. This went on for hours; however, nobody fingered the farmer, and the Soviets didn't shoot anyone. They also searched the houses and barns. In the afternoon, the troops were ordered to leave. They never found out who had killed their drunken comrades. Even under this pressure, not one of the farmers squealed, though they all knew who had killed the Soviets. We were just glad to be out of there.

One of my jobs was to deliver the new or repaired shoes to the customers. I loved that job, because I got tips sometimes. At times, the customers didn't want to pay, and I had to learn to politely prevail for payments. Often they said that they didn't have any change, but these were only excuses not to pay. I always carried change to settle that.

I always loved to deliver shoes to the monastery. It was sometimes difficult to negotiate the hill with eight to 10 pairs of heavy shoes. The priests did not tip; however, sometimes they gave me a piece of candy, and once they gave me a date, which I had never seen before.

One of my most memorable shoe deliveries was to his Highness Duke Lonyai. The monastery gave both the duke and his wife safe haven during the war, and both left all their belongings, including large land holdings like vineyards and castles, in return. They had beautiful and spacious quarters next to the archabbot's. The duke used to have his shoes made in England, but the war stopped that. My father's shop was selected to make shoes for him. They were made out of very soft goatskin. My father didn't know if he was supposed to send a bill to the duke. He told me that if they asked how much it was, I should tell them. A valet took the shoes from me in one of the anterooms and told me to wait, and then he disappeared. While I was waiting, I worried that the duke would reject them.

Finally, his head valet came out and told me that his Highness liked the shoes, that they were just as good as the English shoes, and he would pay the same amount for the shoes my father had made. He paid me in 22k gold, which was worth many times more than what my father would have billed. We were able to live off of it after the war during the horrific inflation. My parents gave some of the gold to my farmer uncle to buy a pair of workhorses, because his had been confiscated during the war. I remember how happy and grateful my uncle was when my father put the gold in his hand.

Another memorable episode in the shoe business concerned a butcher from across the street. He ordered a pair of fancy boots of the finest calf's skin; however, he wouldn't pay for them, and we were left without money for the weekend. Sunday evening, when I went to a dairy to get milk, he was wearing his boots and talking like a big shot with some girls. He stopped me and kept ridiculing me and insulting me. I couldn't take it anymore and told him instead of picking on me, he should pay for the boots. Monday morning, he came through the shop door, threw the boots across the room, and told my father what he could do with them because of his son's big mouth. After school, my father told me never to criticize a customer, and then he took me to the butcher shop and whipped me in front of the butcher. This was a harsh way to learn that the customer was always right.

I always liked examining nature and walking in the woods. When I was about fifteen years old, I discovered mushrooms in a pine forest. They were edible and similar to shitake mushrooms. I took some home, and we ate them without incident. Then I started to pick them by the basketful. This was at a time after WWII when food was hard to come by. Pretty soon, I traded mushrooms for meat with the butcher and bread with the baker, selling some on the side. There was a girl who I was very fond of as well. I took some mushrooms to her family, and they loved them. I immediately offered to take the girl with me the next day to pick some more. I thought I had hit a homerun. I had not only pleased her family, but she had also agreed to come to the woods with me. When I went to get her next day, her mother greeted me at the door, trip ready. I asked her where she was going, and she said she was coming along as a chaperone. This shattered all my fantasies for my trip to the woods with the girl, but that was not all. The next time I went to my secret mushroom field, there were a lot of people picking. Her mother had told her friends where it was. I then learned to be more cautious when it came to women and business.

Duke Lonyai donated his land and palaces to the monastery. One of the palaces was near the Austro-Hungarian border. After WWII, that area was taken away from Hungary and given to Austria. The Benedictine monks didn't want all the paintings, treasures, and antiques to go Austria. Somehow, they hired a truck, which was extremely hard to find, to haul away all the treasures from the palace and deliver them to the monastery before Austria took it over. I had never seen a palace before, so naturally, I wanted to see it. I also wanted to escape to Austria. After the truck was unloaded, I hid behind it, and as it started out, I crawled into the truck bed. It was a cold and bumpy three- or four-hour ride. I saw the palace with all its splendor. Then I asked a shopkeeper the chances of defecting to Austria. He said the Soviets shot everyone who had tried it before. So there I was at age thirteen with no money and no place to go. I asked the trackers if I could hitch a ride with them to Pannonhalma, but they said no. I hid until the truck was loaded, and as it started, I grabbed the back of the truck and pulled myself up and slid in between the antique furniture pieces. It was so cold that I thought I was going to freeze to death. As we got close to the monastery, I jumped off the back when the truck slowed down at a curve. At first, my parents didn't believe my adventure. They were more concerned about my trying to defect.

Chapter IV

Oppressive Hungarian Dictatorship

As the Hungarian Communist dictators were getting more and more control with the Soviet Red Army's help, life was getting worse and worse. Factories and private businesses were confiscated. Farmers' land was either confiscated or taxed to the point where their tax was higher than their income. Then they were forced to give up their land and join the Soviet style collective farming. This type of farming was a total failure in the Soviet Union, and it was a failure in Hungary as well. There were massive arrests, and people or families just disappeared, never to be heard from again. This was how Communists invoked tremendous fear in people. No one was safe, and all feared the knock on the door at 2 o'clock in the morning by the secret police, after which many were never heard from again.

In 1948, the Communist government confiscated all private schools. Our famous Pannonhalma gymnasium was no exception. All teachers were dismissed and replaced by other teachers, many of them Communists. Most of our boarding students left and were replaced by children of high-ranking Communists. They insulted and intimidated the remaining students any way they could. It was not unusual to receive a threat like "I can get you arrested" and "nobody will hear from you again."

Then one of the communist students tried to rape one of our schoolmates. There was a big hearing on the case. Several of us testified. We knew if the case was lost, we were finished. Then all Communist students staged a demonstration and threatened the principal: If the attempted rapist was not found innocent, they would all leave, and the Communist Party would hold the principal responsible for the consequences. The pressure was getting unbearable. We witnesses didn't know if we would all be arrested. Two days later, none of the Communist students came to class. They were all gone. We had won in a Communist system against them!

As the Communist dictators built up the feared secret police, the terror increased. Assembly was forbidden. If more than three or four people gathered, a police permit was required. We had survived all the adversities without being kicked out of school or arrested, and finally, we were going to graduate. We wanted a nice prom to celebrate. We located a dance band that played American music in the city of Györ. They opened up with "Chattanooga Choo Choo," and everyone went wild. As the party got underway, the police chief showed up to check on us. We immediately set him down at a table and brought him a liter of wine and thanked the comrade captain for stopping by. Just as he started work on the wine, the band, which didn't see him, started to play "American Patrol." The comrade captain asked, "What is that pretty song?"

We were mortified, but suddenly, I blurted out, "It is a Russian folk song."

The comrade captain wanted to get the record of it. Shortly thereafter, he thanked us for the wine and moved along. I never had to follow up on the promise of the record, because he disappeared—probably arrested. Then the party really started. All our youthful energy was consumed in dancing the fox trot and swing.

As our gymnasium prospective maturation approached in 1950, the Communist terror intensified. Everybody lived in fear of being arrested. There were many people who were innocent arrested and tortured just to terrorize the population. I was valedictorian, and as I prepared my graduation speech, I had to think about every word I would say, sure not to say anything that could give the Communists an excuse to arrest me. During the summers, I worked as a substitute mailman while our regular mailman took university courses. After graduation, that job was no longer available, so I went to work in a brick factory, mining clay and taking the hot bricks out of the kiln in a wheelbarrow.

I was waiting anxiously for my interview notice from medical school. Finally, I received an open postcard stating that my acceptance had been denied, the reason being "class alien." Those who were unlikely to assimilate into the Communist system, those whose parents had businesses with employees, or those who owned larger farms before they had been confiscated and turned into government-owned entities were listed in this category. At that time, all universities had a Communist-imposed quota system. Of those accepted, regardless of their scholastic achievement, about 60 percent had to be children of industrial workers or collective farmers. The other 30 percent were service-worker families, such as porters, railroad conductors, among others. About 7 percent were from the "intelligentsia," including the children of doctors, lawyers, etc. One-to-three percent were from children of "class aliens," regardless of their scholastic achievements. My father had employees, and it didn't matter how hard he had struggled to get to that point from a starving orphaned boy. In addition, I had gone to a famous parochial school, which was thought to be the best, before the Communists confiscated it from the Benedictine monks. That was why admission had been denied.

My misery was further aggravated by two other events. First, one of my classmates who had come to me before graduation to discuss how unbearable the Communist terror had become. He and I both felt it could not possibly last much longer, because it was our understanding that the Yalta Pact guaranteed free elections and withdrawal of Soviet troops from Hungary by September 1945. It was now 1950, and the Soviet troops were still there. With their help, a most oppressive and cruel Communist dictatorship was established. He thought we should document these Communist atrocities, for when the regime changed, we would have proof of their crimes. They would not be turncoats like so many Nazis, who were first to join the Communist police. I suggested to him that if these documents were to be found by the secret police, we would all be dead. I thought we should just memorize the atrocities.

Then I heard that he was arrested, tortured, and savagely beaten. The people who worked in the forest said that one day the secret police brought him to a location where they had previously buried weapons and dug them up. After repeated savage beatings, they made him admit that he had hidden the guns there and was planning an armed uprising. This was a fabrication of the secret police, because I knew he had no guns. After that, we never heard from him again. His arrest put tremendous fear in me. What if during his torture, the police had extracted my name from him? There was no place for me to hide. The best I could do was to go to Budapest. I moved in with one of my cousins and kept the door locked to the room. I decided if the police were looking for me, I would be able to jump out of the fourth floor window before they could break down the door, and they wouldn't be able to catch me alive. Slowly, my fear lifted like a fog.

The other problem I faced was the draft or rather conscription to the Red Army. Because I was classified as a "class alien," I would be assigned to a penalty unit, meaning that during peace time, I would be cleaning latrines and doing other such menial jobs. In wartime, class aliens would be placed in front of the real troops so that the enemy would shoot at them first. In this way, the regular troops could better determine where the shooting was coming from and therefore, be more successful targeting and hitting the opposing army.

Chapter V

College Years and Coping with the Red Dictator

I arrived in Budapest for the first time in my life. When I was a kid, my father used to go to Budapest to learn about new shoe fashions. He always came back with exciting presents and stories about the zoo, monuments, parks, etc. We thought Budapest was magic.

Now, here I was in Budapest, poor, worried about whether or not I was going to be arrested because of my friend's connection, not in school, without a job and with hardly any money. I walked from the railroad station to my cousin's rented room to save money on streetcar fares, especially because I was also afraid that I would get lost on the streetcars. My first priority was trying to get accepted to a university to escape the draft to the Red Army. I had the idea of getting a job, any job—janitor, coal carrier, or anything in the medical school. That way, if somebody dropped out or flunked out, I might be able to take his place. The various departments or clinics were in separate buildings in the medical school. They all said the same thing: "With your grades, you are overqualified, and you should be in the university as a student." I agreed, but none of them could help me get in. I was running out of money and becoming more desperate with each passing moment.

As a final effort, I went to the university admission section of the Ministry of Education to plead my case. Bollo Bertalan, the counselor, was sympathetic but said there was nothing he could do for me. I kept begging, and he said I had to leave or he would have the police take me out and arrest me. I told him I was out of money and food, and I couldn't pay my rent. At least in jail they would feed me and give me a cot. This must have touched a note in his heart, because he picked up the phone then. At this point I didn't know if he was calling the police or what. But it turned out that he talked to the director of admissions at the Therapeutic Pedagogy College. When he hung up, he asked me if I would like to be a student there and become a therapeutic pedagogist. I told him I would really like that, although I didn't have a clue what a therapeutic pedagogist was; however, it would mean that I could avoid being drafted into the Red Army. He then told me to report to the admissions director of that college.

I remembered that one of my gymnasium classmates who was from Budapest once mentioned that they had a family friend who was a therapeutic pedagogist. On the way to my college interview, I looked up my friend and found out that a therapeutic pedagogist could be a deaf, special education or blind school teacher or even a physical or speech therapist, depending on his or her major. My interviewer was pleasant looking and didn't dwell on my parochial school education or being a "class alien." Then she said that I should meet the president of the college, a very warm physician and renowned authority on mental retardation. He said they would be glad to have me, and if I made good grades, they would give me a cash scholarship. Suddenly, I was a college student and on cloud nine. I made very good grades, enjoyed school immensely, and got the scholarship.

My fear of being arrested for the Sylvester Farkas affair slowly eased. It appeared that the secret police could not get Sylvester to name his conspirators.

In college, my classmates came from various backgrounds. I suspected some were "class aliens" like me, but we didn't dare discuss our backgrounds with each other. One of my classmates was a shoemaker and had never gone to high school, but his uncle was a counselor in the admissions department at the Ministry of Education, so he put him in this college. He had a lot of problems with the subjects, particularly with Latin terms. Those terms were easy for me, because I had taken eight years of Latin in gymnasium. I studied with him and helped him. He tried very hard and made passing grades. Later, he excelled and eventually became president of the college. He didn't forget my help. Near the completion of our first year, we all were desperately looking for summer jobs. He asked me if I would like to work in the University Admissions department of the Ministry of Education. I told him I didn't have a chance for such a good job, but he said that he had already talked to his uncle and that they needed temporary workers in the summer, so he would be able to get both of us jobs there.

In less than a year, I had gone from an unemployed university reject to a top student with a scholarship and a summer job.

At college, the Communist Party secretary of our class came from a heavy industrial background. He had trouble keeping up academically and came to me for help, so I helped him, too. During the school year, the president of the college got me a part-time job tutoring retarded children who were part of his private practice. He had a hideaway private office, and he let me use it for work. This all happened when the Communist terror was at the highest point. There were massive arrests, deportations, confiscations of private properties, fake trials, and even executions. However, in our little college, I was happy and doing well.

But one had to constantly be aware of the oppressive ways of the Communists. If they found out that someone was a churchgoer, he was fired from his job. Similarly, if the Communists found out that I attended mass, they would use that as a reason to end my college studies. To try to avoid this awful outcome, I would go to a different church every Sunday. As I approached the church, I would use a handkerchief to supposedly blow my nose and cover my face to avoid being recognized. I never used the same door to enter and exit, and thankfully, I was never caught.

One of our major clinical rotations was the National Rehabilitation Institute, where we learned physical therapy. The director was a renowned rehabilitation physician who was a Jew and had escaped to Switzerland before the Nazis could intern him in a ghetto. He had became very famous in Switzerland, and after the war, he came back to visit Hungary; however, the Communists would not let him leave now.

Everybody was required to write a thesis and defend it. My thesis was on therapeutic pedagogy and how the physically handicapped people, such as polio victims, could be considered pseudo-retarded, because the affected children were not exposed to the learning opportunities and life experiences that their able peers were. The study was well received, and it addressed Professor Petö's concerns regarding the pseudo retardation of these unfortunate patients. Later on, it was decided that because most of the children could not attend regular schools, there should be a school established at the National Institute of Rehabilitation with emphasis on the correction of pseudo-retardation in order to raise their learning to par with their non-handicapped peers.

During my college years, I still had a fervent desire to be a physician. We learned that in the Soviet Union there were a few physician/therapeutic pedagogists called defectologists. I approached the president of the college about the idea and asked for his help to get me into medical school. He was lukewarm at first. I needed more allies, people with more pull for his support. I approached the Communist Party secretary at the college about the idea. He was a classmate, whom I had helped with Latin, and he was an older man with a family. During our discussion, he told me that he wanted to go to medical school as well, but he could not attend because his grades were very poor.

I then approached another classmate who was also a class alien reject from medical school like me. I also knew that he was a nephew of the president of the college, which was kept secret. The three of us went back to the president with the idea of going to medical school after graduation. Now the president approved our plan and offered his help. He was not a Communist, but he was a well-respected, leading authority on mental retardation. Many of his private patients were from the higher echelons of the state, which provided him with good connections.

Now, I was hoping and praying that my class alien status would not keep me from being accepted to medical school. This was my best chance to become a doctor. I decided not to mention that my parents had employees in my application, because that would keep me out of medical school. One question on the application asked if I had any friends or relatives who lived in a Western country (meaning the free world). I decided to answer "no," even though one of my cousins, a talented young painter, defected to Italy. By answering this way, I avoided rejection. I took the risk, knowing that if the medical school administration found out about it, I would be kicked out.

During the next three and a half years while I was a medical student, whenever I was notified to report to administration, I became terrified, because I never knew if they had discovered the omissions on my application. Fortunately, these omissions never came up during any of my visits to administration. My good luck had started.

Our recommendations from our college president were so influential that we did not even have to have interviews or entry exams. We were just notified that we had been admitted. Things were starting to look good. Our college president still referred pupils to me for private tutoring, so I had an income as well. I also had a scholarship. I could support myself and savored every minute of my medical studies.

In the meantime, the cold Budapest winter came, but I did not have a warm coat. My parents sent me money for one. My younger sisters were taking piano lessons, but they had no piano to practice on. In one of the consignment shops I used to go to, they told me that someone was trying to sell an upright piano. Eventually, I got that piano with the money my parents sent for the winter coat along with several sides of bacon and a ham. As for the winter coat, I picked up a used one in the consignment shop. Consequently, one of my sisters became a very good piano player. She then switched to the organ and supplemented her income by playing and singing at weddings.

Things were going relatively well. I could only afford to rent a small, unheated room; however, I was on my way to becoming a doctor, and that was all that mattered. In the late fall of my second year of medical school, my right ankle and knee started to hurt as I got on the streetcar to go to tutor a pupil. By the time I had reached his apartment, I could barely walk. When I finished tutoring, I could barely stand up. They loaned me a cane, and I dragged myself to the streetcar station and went straight to an outpatient clinic. There, the doctors told me I was seriously ill and needed to be hospitalized immediately. However, all the hospitals were full, and there were no beds available at the time. Finally, they found a hospital bed for me on the other side of the city. There was no transportation, and I was told I had to get there on my own despite the fact that I could not walk. My brother's landlady had a phone, and the nurse let me use the clinic's phone to call him. My brother received the message and came and dragged me from streetcar to streetcar. He even had to carry me on his back the one-half-mile walk from the streetcar station to the hospital. It was a cold night, and by the time we got to the hospital, his clothes were soaked with perspiration. Finally, I was admitted to a 12-bed ward. Most of the other wards were thirty to 36 beds.

After numerous consultations, my diagnosis was either Reiter's syndrome or severe arthritis. I was given aspirin in such high doses that I lost my hearing. I was very well treated, probably because I was a medical student. One of my doctors told me that my prognosis was very poor unless I could get hold of some aureomycin and use it. This drug was manufactured in the Western countries, and none was available in Hungary or the Soviet Union. If someone was found to have aureomycin or any other free world product, there would undoubtedly be a penalty.

Fortunately, I had many visitors, and I mentioned my need for aureomycin to some. Miracles happen! One day, one of my colleagues from the National Institute of Rehabilitation came by and put two bottles of aureomycin by my blanket and said that Professor Petö wished me a fast recovery.

Unfortunately, the aureomycin did not help, and I was in bed for a long illness. Because I was always lying in bed, I rapidly started to lose my muscle mass, and I got to the point where I could barely get out of bed. In the meantime, I developed a heart murmur, and the doctors changed my diagnosis to rheumatic fever with carditis. My hopes of being let out of the hospital and taking my semester closing exams evaporated.

Janos Hideg, one of my classmates, persuaded the professors to come to the hospital and give my oral exams there. Now I had to study really hard, but I was limited by the light available, because the lights were turned off after certain hours. My brother gave me a small mushroom lamp, which I clipped it to my books, covering my head and books with my blanket to study late at night. The lamp generated a lot of heat under the blanket, but I did not matter, because I needed to get ready for my exams. If I passed, I wouldn't lose a whole semester. I made A's and B's, but there was a problem with military science. Our military science professor did not want to bring a rifle to my hospital ward, but eventually, he agreed to bring his service pistol to demonstrate my knowledge of guns and their maintenance.

I was a patient in the hospital for five months. I kept studying so that I would pass my year-end exams and avoid losing the school year. After I was discharged, I walked with a cane for a long time, but I was recovering nonetheless. To this day, my right quadricep is weaker than the left, but it does not stop me from playing tennis and exercising.

Chapter VI

New Adventures and the Slipping Red Dictators

The next year, the school became a reality at the National Institute of Rehabilitation, and Dr. Petö asked me to take the assistant principal position. When I told him that my medical school studies might interfere with my duties as assistant principal, he said, "I've watched you, and you can handle both. The patients are here all the time and you can handle your class teaching in the afternoon and your administrative duties at night." This was the most exciting assignment—new school where our physically limited pupils could mentally develop.

When spring came, my landlady let me know that she was putting their winter clothes into storage. There was extra room, and she said that she would be happy to put my winter clothes there as well at no cost. I gave her all my warm clothes and offered to help take them to the storage area. She said it wouldn't be necessary, because they were not taking them yet.

In the fall, I wanted to go home for my parents' wedding anniversary on October 10th. However, the weather was turning cold, so I asked my landlady around mid September if I could get my winter clothes from storage. She said that I could but that she needed a couple of days. The days became weeks, and every time I asked, she told me I would have them tomorrow. October 10th was quickly approaching, and I eventually told her I had to have them by the next day. She told me that my clothes would be waiting for me in my room after school before I went to the rail station. There were no clothes in my room the next day. She then told me that she would get them right away and that I could take off my summer clothes in my room and she would hand me my winter clothes to save time changing. While I was waiting, partially undressed, she came in my room and hugged me very passionately. I pushed her away and demanded my clothes. She told me they would be waiting when I came back.

When I got home to my parents house, they were very concerned that I would catch pneumonia, being so lightly dressed. None of my clothes were waiting for me when I returned to Budapest. I talked to my landlord, and he told me that my clothes were gone for good. When I asked him to replace them, he told me that they had no money for that.

I decided to report the theft to the police, but before I did that, I talked to my college party secretary. He also recommended that I go to the police. I still did not know who my landlord was, and with his influence, he could get me arrested. I asked the party secretary to use his influence to look for me should I suddenly disappear.

My first visit to the police station was encouraging. A lieutenant wrote up my case and told me to come back in two days. When I came back, the treatment was not as kind. They told me that the case had been taken over by the major and that I needed to see him. The major told me to come back at nine o'clock the next night. When I got home, my landlord told me that I should have never gone to the police, because it would hurt everybody, including me.

The next day at nine in the evening when I went to see the major, he was quite unfriendly. He warned me that if their investigation found that I libeled these prominent citizens or was trying to extort money from them, I could face a long prison sentence. He told me to think it over for a few days. If I dropped the charges, he said, that would be the end of it. If I did not, they would have to "investigate," but it seemed to him that I accused some innocent citizens. As I left, I was scared to death, so I went back to my college party secretary and told him what happened.

In the meantime, I was scared to go back to my room by myself. The party secretary told me to go see the Budapest police chief the next morning, that he was expecting me. I didn't know at the time that they were from the same town and both were idealistic Communists who could not stand the Soviet imposed dictatorship. During the uprising, the police chief was on the freedom fighters' side and ordered the police to support us.

After I told the police chief my story, he made several phone calls and told me to go back to the district police and see a certain major who would handle my case from then on. He also told me not to worry about my personal safety, which was a great relief. The new major was an older man who was once a factory worker that had turned into a police officer.

He told me I didn't need to worry about libels, and they would pick the landlady up around two o'clock for questioning. He said that she had done this before.

The feared knock at the door came as promised. I wasn't sure if they were going to arrest her or me, but it turned out they were there for her. My landlord was quite agitated and offered a very small settlement if I dropped the charges immediately. It was way too small, and I wanted replacement costs. I realized the danger of his police connections, but I stood firm.

After the police interrogated her, she was released. Her husband offered again a slightly different but still inadequate settlement that would be paid over a longer period of time. I rejected that as well. The landlord made a threat and told me that I would regret rejecting his offer. I knew I had to get out of there as fast as possible.

My brother heard about a shared room rental, and I told him I would take it. The two of us could easily carry my belongings. After we got there, I realized it was not exactly an ideal place, but it was better than sleeping at the railroad station. In the room, there were only pegs to hang my clothes along with the landlady's. I would have to share the bathroom with others. The "kitchen" was on one side of the wall, and there was not enough light anywhere to read or study. I expected a little room or cubbyhole where I would sleep. When I asked where I was to sleep, she pointed to a sofa at the foot of a double bed.

Later, after the light was turned off, there was a knock on the door. Luckily, it was not the police. It was a man who had come to get the pleasures of the flesh. Without hesitation, they went ahead with the usual sounds of such activity. When he finished, he paid and left. I knew then that my new landlady was a hooker. This situation was hardly conducive for study or sleep. I immediately started to look for another place. Luckily, I found another "maid's" room in Buda, which was also unheated. It was small, but at least it had a window.

In the meantime I got nowhere with my thieving landlady and landlord. The father of one of my classmates was an attorney who was one of the few lawyers who still had a small private practice. He offered to take my case pro bono. In no time at all, he got me a good settlement—not replacement value but good enough. Now I could look for replacement clothes.

One of my acquaintances from the Ministry of Culture was one of the people I asked to look for me should I disappear or get arrested. I went to thank him for his help, and he gave me the address of a tailor shop and told me to tell the manager that he had sent me. This shop was not open to the public and made high quality merchandise for government VIPs. The prices were so low that I got a new winter coat for a fraction of what I would have paid for a used coat in a consignment shop. A classmate of mine told me about a tailor shop that had a lot of western diplomats as customers. He told me that if I advised the owner on how his son could get into medical school, he would be very grateful. It turned out that his son couldn't get accepted because he was rejected as a "class alien" like me. I told him that I didn't have any pull, but I would think about it. After a while, it hit me that the Communist Party secretary, who was a medical student and had also sent me to the chief of police, always liked nice clothes but could never afford them. Of course, I told the tailor about him. I thought maybe something could be worked out between them, and I told him that I wanted to bring him by. The following September, the party secretary was wearing a beautiful, tailor-made, navy blue suit, and the tailor's son was a freshman in medical school. I also got a nice sport coat out of it, which I happened to wear when I escaped. I still have it today, and it still almost fits me.

Chapter VII

Uprising

In 1956, the iron grip of the Communist dictatorship seemed to loosen a bit. This was three years after Stalin's death. What I did not know at that time was that the Hungarian Communist dictators were more cruel and repressive than Stalin and a delegation of Soviet Politburo members came to Hungary ordering the Communist puppets to loosen their grip and give a little slack to Hungarians. Once this loosening started, there was no way to stop it. Some of the writers openly demanded more freedom, which was very refreshing to see. In the spring, hope was in the air. In 1956, there was a medical student exchange with Romania, which was another Communist dictatorship, but it was the first time students were allowed to go to another country. I was selected to be one of the exchange students.

In the fall, people seemed to be less afraid of the regime. One could hear open criticism of the Communist system. In October, there was a class meeting scheduled for my medical school class of 320 students. These meetings were Communist puppet shows. Usually, a Communist or Communist sympathizer was nominated to conduct the meeting. The questions were prearranged and given out to party members. Most of them praised the Communist system. The person who presided over the meeting was supposed to recognize only these people; however, this meeting was very different. When a party-faithful person was nominated to preside over the meeting, many of my classmates started shouting that they had another nominee, Laszlo Makk, and demanded to put the nominees up for a vote. Consequently, I was voted to conduct the meeting. When I walked to the podium, a Communist behind me whispered in my ear that he would tell me whom I should let ask questions. I decided to ignore his directive, and

I gave everybody a chance to speak up or ask questions. The comments were more and more critical of the system. They ranged from our military camp cruelties of uneducated officers to the reasons we had to take up 12 hours a week with obligatory Marxism and Leninist doctrine, military science, and Russian language studies when none of these helped to make us better doctors.

After the meeting, the Communist behind me was livid, and as he was leaving, he whispered to me, "Makk, you will pay for this." Then I realized that this was the first free class meeting where my classmates could express their grievances. It was an exhilarating feeling, but it did not last very long. That evening, several of my classmates came by to advise me not to come to school the next day, because I would probably be kicked out. My life's ambition to be a doctor was in danger again. The next day, demonstrations were planned by all the universities at two o'clock to express solidarity with the polish workers who had an uprising against the Communist system. I just had to be there. Our medical school was part of the University of Medical Sciences, which included dental school, nursing school, and teaching hospitals. It occupied a city block that was surrounded by a tall, wrought iron fence.

When I got to the medical school, nobody was in class. Everyone was on the grounds, milling around and waiting to demonstrate. At the last minute, the president of the university forbade anyone from demonstrating and had the gates locked. He personally supervised the locking of the biggest gate. This got the crowd excited, and students who were nearby pushed him into the guardhouse. They somehow took the iron gate off of its hinges then. It hit the ground with a deafening noise. We poured out to the street, and we were no longer afraid of the Communist beasts. We demanded freedom, which was formulated into a 16-point demand declaration.

As we filled the street, chanting our demands and singing Hungarian songs, the windows of the apartment houses opened, and people started throwing flowers at us. Then they started to wave the Hungarian flag with the Communist hammer and sickle symbols cut out of the centers. As we marched by a flower shop, the people who worked there threw all their flowers at us and joined the demonstration. People from offices and shops joined us, too. These were some of the most exhilarating moments of my life. After 10 years of suffering, subjugation, and fear, we showed our defiance to the Soviet dictators and their Hungarian puppets.

The radio building was a city block long. In addition to radio, it was a Communist propaganda center. We wanted to broadcast our sixteen demands, but we were refused. They wouldn't even let our representatives in the building. After a while, we were ordered to disperse. We were quite crowded by now, because droves of people had joined our ranks. We were, however, very peaceful. After a while, some of the windows opened at the radio building, and we could see gun barrels sticking out. Suddenly, the secret police inside the building opened fire on us. A classmate whom I saw fall was only about 10 feet from me. I rushed over to him and dragged him to the nearest apartment entrance for protection and first aid. He died a few minutes later from a gunshot wound to his chest.

The firing continued, but we kept demonstrating more defiantly. Then a truckload of Hungarian Secret Police (AVO) reinforcements drove into the crowd. Someone ingeniously set fire to the truck, and it exploded, killing most of the AVOs. A while later, other trucks appeared. They were loaded with factory workers, guns, and ammunition. These workers were from an ammunition factory. When they heard about our struggle, they disobeyed their superiors, loaded up the company trucks with guns and ammunition, and joined us. They kept throwing guns and ammunition from the trucks to whoever wanted them. Now there was return fire, and our demonstration turned into a fight for our freedom.

Word came that there would be a demonstration in front of the parliament building. We decided to go there next. The huge square was packed. At first, one of the Communist dictators tried to calm us down, asking that we disperse. We booed him until he left from the parliament building window. Then another moderate Communist who was more Hungarian than Soviet tried to address us through the loudspeaker. When he addressed us as comrades, he was booed. When he addressed us as ladies and gentlemen, a big cheer rose up from the huge crowd. Somehow, this was the first sign of victory, and we hoped that we would never again be addressed as comrades.

After the parliament demonstration, we decided to go to the national printing plant to request they print our sixteen demands. They did it right away. Later, they were distributed. Somebody came from the radio building and told us the gun battle was still going on there. Because there were no doctors there and I knew triage from taking a military science class, I went right away and set up a triage station in the lobby of a nearby building. Before I realized it, I was in charge. Volunteers brought medicine, bandages, and food.

It was important work, but the horrors were hard to handle. I can still see the face of an 18-year-old student who was mortally wounded and died in my arms. I still remember the three people who were brought in, gesturing that they could not hear. We suspected they were secret policemen who had taken the insignias off their uniforms but hadn't had a chance to change into civilian clothes. We sat them on chairs facing the same direction. I went behind them and in a loud voice told one of the freedom fighters that if they didn't talk, then maybe we should shoot them. Their facial expressions changed. I slammed the door behind them, and they all jumped, so I knew they were not deaf. I told the freedom fighters that they didn't belong there, pretending to be deaf and taking up valuable time at the triage station. On the way out, one of the fighters assured me that I didn't have to worry about them ever coming back again.

One morning, a convoy of army trucks drove up along the radio building loaded with armed soldiers who were cadets of the military academy. We knew that if they attacked us, we wouldn't last long. We would all wind up dead, wounded, or captured in no time at all. Since I had a white doctor's coat, our group asked me to carry a white sheet on a stick and walk up to the lead truck to find out what they were up to. They assured me that I would be covered if they started shooting at me. Of course, I was scared, because either the secret police in the radio building or the cadets from the convoy could open fire on me at any moment, but it never occurred to me not to proceed.

As I approached the commander's truck, he ordered me to stop and identify myself. "Laszlo Makk, fourth-year medical student," I said. The commander was standing on the passenger side of the command truck, his upper torso stuck out through the open hatch. He ordered me to get up on the hood of the truck and state why I had come. I explained our sixteen points to him and told him that we were students who wanted a better life and more freedom for all Hungarians, that none of us were criminals or hooligans and that the only criminals were the secret police, who had killed innocent demonstrators. I also told him that the radio building was full of secret police who had started the shooting, killing many unarmed Hungarians. I knew many of the soldiers and policemen hated the secret police for their privileged status. I also knew many Hungarian soldiers hated the Soviets, because they treated them like serfs. After several probing questions, he told me to go back to where I had come from. I knew if he was a devout pro-Soviet Communist, they could just shoot me in the back, but I kept on walking.

In the meantime, we thought we heard someone from the radio building using a loudspeaker to order an attack. After a few tense minutes, the major in charge of the cadets gave the order to attack—not us but the secret police inside the radio building. We were overjoyed. After a short attack, the secret police gave up. When they came out of the building, the ones in front had their hands up, but a few in the back had their machine guns at the ready. When the ones gesturing surrender stepped aside, the others in the back opened fire. Many of the Communist secret police were killed, and the cadets who were now on our side captured the rest.

I had many magnificent moments in my life. My proudest was when I walked over to the commander. At that time I just did what I had to do without hesitating. Unfortunately, there was a sad end to this. Two days later, I was riding in an ambulance on a boulevard downtown at around nine o'clock in the evening. Suddenly, we heard the treads of Soviet tanks. We needed a hiding place immediately. We saw a movie theater with a partially destroyed front. The driver backed the ambulance in there, and we hid among the seats. We heard some groaning. We located the man in the dark. It was the brave major who had ordered his cadets to come to our side. Outside, the Soviet tanks rolled down the boulevard, firing their cannons at buildings randomly, hitting a hospital and a church. With only a match for a light, we found that the brave major's right leg was almost completely blown off. All I could do was use his belt to make a tourniquet for his thigh. After the tanks left, I told the ambulance driver to take him to the safest hospital and to tell the doctors if he survived, he needed to escape to the west. If the Communists got a hold of him, he would be executed immediately for coming to our side.

A few days later, all major fighting in Budapest ceased. The Soviet troops withdrew from Budapest, and the Soviet ambassador started to negotiate the withdrawal of all Soviet troops from Hungary. These were glorious moments, but they did not last very long.

The country was free at last, and the Soviets were withdrawing. (They were supposed to withdraw in September 1945.) The country was elated. We had a temporary government, and Soviet Ambassador Andropov negotiated the withdrawal. On Saturday, November 3rd, it was announced that an agreement had been reached and that all Soviet troops would leave. The Hungarians were to rebuild all destroyed Soviet war monuments, and they would organize farewell parties for the departing Soviet troops. It sounded too good to be true. It was also announced that a Hungarian delegation would go to the Soviet headquarters that night to sign the final document for withdraw.

We had won the uprising! We were free and proud to live in a free Hungary. The major credit for our victory should go to Colonel Paul Malleter, a Hungarian on the privileged Soviet side and a young partisan during WWII. The Soviet Army parachuted him several times behind the German frontier in Hungary, where he conducted sabotage and organized resistances. After the war, he was a rising star in the Soviet-controlled Hungarian People's Army. He joined the Communist Party and had training in the Soviet war college for senior officers. He learned Russian and was fast rising to the top in the officer's corps. During those times, I was sure he got disillusioned with how the Soviets were taking away Hungary's natural resources, how they treated the people like serfs, and how the pro-Moscow Hungarian Communist puppets doled out such cruelty. These puppets arrested tens of thousands of innocent Hungarians and confiscated their homes and properties. They separated families and deported them to hard labor camps in secret locations. They cut off all their communications, and no one knew where their spouses or children were.

However, Malleter was a Hungarian first despite his privileged status under the Communist regime. He became the military leader of the uprising against the Soviet Army and Communist dictators. He won battle after battle against superior Soviet Army units. His headquarters were based in an old Hungarian Army fortress built in the 1700s, one with very thick brick walls. The Soviets kept coming to try to take him and his soldiers, many of them teenagers. I remember walking near the fortress and seeing approximately 10 Soviet tanks all burned up with dead soldiers leaning out of the turrets or laying around the burned out tanks. The Molotov cocktail was invented here, and it became very effective against Soviet tanks. It consisted of a flammable liquid, and if it was thrown against the rear grill of the tank, it caught fire right away.

On November 3rd, the guns were silent. Soviet troops were withdrawing from Budapest, and the city was celebrating. A high-level Hungarian delegation was invited to the Soviet headquarters to partake in a banquet and to sign the Soviet troop's withdrawal treaty. Most leaders of the uprising, including Malleter, were in the delegation.

At four o'clock on Sunday morning, November 4th, Budapest was awakened by the deafening roar of the firing of hundreds of Soviet cannons and tanks. I was in Buda, and as I looked over to Pest, it was lit up by fires. Then the first prime minister of free Hungary came on the radio begging for help from the free world. Without provocation, the Soviet Army had attacked Hungary with overwhelming force. His pleas fell on deaf ears. Later, I found out that Chester Boland, the U.S. ambassador to Moscow in Eisenhower's administration, indicated to the Soviets that the United States considered Hungary part of the Soviet sphere of interest. In other words, the United States would not interfere. We also found out later that when the Hungarian delegation reached the Soviet Army headquarters, a detail of heavily armed Soviet secret police appeared ready to fire and arrested the entire delegation.

My immediate goal was getting to my triage station in Pest. I started out and tried several streets, but they were all blocked by Soviet tanks. Finally, I got to a street that looked okay. As I started to walk down it, I heard the tracks of a moving Soviet tank. Then, there it was, aiming a cannon. I jumped between two gateposts of a yard, hoping they hadn't seen me. The tank headed right up the street where I was hiding. I don't know how I had the strength to push the locked gate open, but I did it and laid down on the side of the house motionless. As the tank went by, the ground shook. They apparently hadn't seen me. Otherwise I would have been blown to pieces.

There was no way to get to Pest, because all the bridges between Buda and Pest had now been blocked by Soviet tanks. Streetcars and buses were out of order, and pedestrian traffic was halted and turned back by Soviet soldiers. The motorized Soviet units kept pouring into Budapest. I ran into an ambulance team who needed a medic, and I joined them immediately. We picked up the wounded and helped deliver medical supplies. It was dangerous, because the Soviets were shooting at anything that was moving. The freedom fighters were shooting as well, because the secret police sometimes used ambulances to transport reinforcements, ammunition, and guns. More than forty years later at our Hungarian high school reunion, one of my classmates said, "Makk, you are lucky to be alive, because I almost killed you during the uprising."

I asked him why, and he said that he had been a freedom fighter with a telescopic rifle. He saw our ambulance and thought that we were helping the secret police, so he was going to take out the driver. As he aimed next to the driver, he saw a familiar face. He realized it was me, so he didn't take the shot.

On another occasion, we were going to pick up a wounded freedom fighter when we ran into a Soviet Army blockade. They took us out of the ambulance and put us on a truck. We were driven to a house that looked residential, but it turned out to be a Soviet Army command post with a captain in charge who spoke fluent Hungarian. I was separated from the driver during the interrogation. The Soviet captain was very poised and polite. He started out by offering me a cigarette and asking me to help him understand how the revolution had started. I told him I didn't know anything about that and that I was working at the National Institute of Rehabilitation, where we even had patients referred to us from Moscow. The interrogation went on for hours. Then, without saying anything, they put me in an empty, dark room. I sat on the floor, and every time I heard the Soviets talking, I didn't know if they would come in and beat me, shoot me, or bring me food. In the morning, they put me on a truck and made my driver show us the way to the National Institute of Rehabilitation. They noticed everyone greeting me warmly. They wanted to see the Russian patients, but they were no longer there, because the Soviet embassy had picked them up as soon as the uprising had started. Now I was in trouble. No Russian patients. In desperation, I told one of the doctors to get me their medical records. When they were brought to us, the Soviet captain could see them, including their referral letters from Moscow in Russian. They were satisfied, wished us well, and then left. The Lord must have been looking after us then.

As time went by, I found that I couldn't locate my brother, Tibor. His landlady told me that she had not seen him for days. I thought about trying to escape while the western border of the Iron Curtain was still open. I figured if my brother was no longer around, I would be the only one who could take care of my parents when they got old, which was the custom in Hungary at that time.

Our helper in the ambulance was a Hungarian who had a very unusual background. He told us he was a spy for the British. When the British got suspicious that the Hungarian authorities might get him, they smuggled him out to Vienna, where he waited to be transferred to England. Then four men jumped him, and he woke up in a Hungarian prison. After much beating and torture, they told him he would probably be executed. His only hope was that he might be exchanged for a Soviet or Hungarian spy. During the uprising, all political prisoners regained their freedom. He was one of those now-free political prisoners, and he was now helping us where he could.

Then, we had another close call. We were trying to get a wounded student from his engineering university dormitory to a hospital. He had an open fracture of his ankle. We parked the ambulance in the alley a block away. We were giving him pain medication and putting him on the stretcher when Soviet troops poured in the front door just as we were ready to leave. They were trying to take over the dormitory. A big fight erupted in the staircase. The freedom fighters started throwing hand grenades and firing their guns. The Soviets fired back, and we tried to figure out how to get to the attic to hide our wounded freedom fighter. Instead, we quickly took the wounded student down the back staircase to our ambulance and successfully delivered him to a nearby hospital.

As time went by, the Soviets kept taking Budapest. Fighting and resistance became only sporadic. I went home to see my parents. They were extremely proud and very concerned for my safety. When I left, we knew we might never see each other again. No words were spoken. After I had shared tears with my parents, I knew it was time for me to move on, because there was no time for emotions. My father then reached into his pocket and pulled out his pocketknife. He gave it to me and said I shouldn't hesitate to use it if the occasion came.

Time passed, and the iron grip of the Soviet forces was getting tighter. One morning, there were posters posted all over Budapest. The commanding general of the Soviet forces imposed a curfew, and anyone caught breaking it was shot. There were checkpoints all over Budapest. We soon realized our helper, the spy, needed to get out of Hungary, because if the secret police caught him, he would be executed right away. He had no personal identification booklet, which would usually contain a picture and pages of personal information. He needed one badly, because he would not be able to get through any of the many checkpoints without it. Fortunately, someone got one for him, but we didn't know if it had been stolen or taken from a corpse. Either way, it didn't matter. Everything matched except the hair and eyebrows. Our spy was blond, and the person whose personal identification booklet we had had dark hair. We quickly realized we needed to dye his hair to match the picture.

My dermatology professor was very active in the uprising. Luckily, his wife was a cosmetologist. We went to their apartment in the ambulance, and she was able to dye our spy's hair and eyebrows dark. We made him memorize all the data in the identification booklet, but it seemed like it was taking him forever to commit the information to memory. The driver and I guarded our ambulance, which we had hidden in a destroyed store. By the time his new hair and eyebrows had been finished, it was already getting dark, and the curfew was still in effect: however, we needed to get him across the St. Elizabeth Bridge that crosses over the Danube to Buda so that he could start his escape to the west. Our ambulance was marked with big red crosses, so we decided to take a chance driving him over. There was a checkpoint on each side of the bridge. We were stopped, and Soviet soldiers opened the car door and yanked us out of the ambulance by our coats. We were very scared, but we tried to smile and act like we were not afraid. We explained that we needed to get food to a hospital filled with crippled children. They found no guns in the ambulance, and after they spoke to a superior, they ordered us back to the ambulance and said *davay*, meaning something like "get quick." And we did. As we approached the Buda side of the bridge, another Soviet tank blocked our way. They lowered the cannon until we could see the dark hole of the barrel. This was the scariest thing I had ever seen, but they did not open fire. It turned out that there was another checkpoint by the tank, and after we told them the same story, we were let through.

We broke up for the night, and there came a bittersweet moment when we wished each other good luck the next morning. Then the driver told me that he was going to escape, too. We were like brothers, but it was over now.

Chapter VIII

Escape

All universities were still closed, including the medical school. I stopped by the medical school to see what news there was. I ran into my classmate and friend, Janos Hideg. He said, "Makk, what are you doing here? You are on the wanted list."

I asked him if he was kidding. He told me to go and talk to the porter. The porter told me that a group of Soviet soldiers and secret police came and made him open the administration offices the day before. They were looking for the addresses of some students, and my name was on the list. He said, "If I were you, I would not be around here."

That was it. I needed to get out before they came and put me in a boxcar to a Siberian prisoner camp. I had nothing but a briefcase with me. I went to see Professor Petö at his apartment and told him what had happened. He told me to escape right away and write them if I could. Then he gave me two addresses to memorize and said that if I got as far as Vienna, they would help me.

When I returned home, my landlady said that the housemaster (similar to a doorman but with more authority) wanted to know when I would be home. I didn't know whether he wanted to ask a medical question, wanted to warn me about the secret police looking for me, or had been instructed by the secret police to report when I had returned home. I left without taking anything. I spent the night with friends, planning my escape for the next morning.

My friend told me her next-door neighbors, a young architect and his wife, were trying to escape as well. We got together, and I told them that I would need to start early in the morning. They said they would like to escape with me, but they couldn't leave until they had located her father. He was not at his place of living, and the neighbors thought he was volunteering in a hospital. They had a phone and knew the name of the hospital where he had last been seen.

After several phone calls, someone from the hospital told us that he was helping unload supplies and had been crushed to death between the loading dock and a truck. He was taken with the dead freedom fighters and was probably buried in an unmarked grave. It was tragic news, but the couple decided to join me in the early morning escape attempt.

We couldn't go to the railroad station in Budapest, because we had heard the Soviet troops frequently surrounded the station and herded everyone except children and the elderly into boxcars to be deported to Siberia. We wanted to reach a suburban station where we could disperse more easily if Soviet troops appeared. On the way, we had to go through a large square called Szenas Square. This had been the site of many fierce battles between the freedom fighters and Soviet forces. Nearly all the buildings were heavily damaged. There was still a large tank detail guarding the square as well, but we had to get through a corner of it. It was still dark, and the curfew was in effect. We got down on hands and knees to avoid detection and crawled very slowly. Finally, we got through, and we could proceed to the Kelenfold Railroad Station.

A cold wind was blowing with occasional sleet. We got to Kelenfold Station but stayed away from the station so that we could disperse in case the Soviet troops appeared. I ran into Laszlo Csele, a friend who had a PhD in economics. I asked him where he was going, and he said, "The same place you are."

I told him I was going home to help my parents butcher the hog. He just smiled and said that we should go together. It seemed to take forever for the train to arrive. As we were waiting, a railroad worker walked by and said, "Hey, Doc."

I did not recall seeing him before, and I asked him if we had met. He said no, but he was a fighter at the radio building and saw me go up to the commander of cadets to bring them to our side. I didn't know if he was real or was a secret policeman trying to make me admit that I had been there. I told him it had to be someone else, perhaps my brother. He leaned closer and said, "When the Soviets raid a train, they never check the locomotive." He then told me that he could get me and my friends up there so we would be safe until the train got close to the border. I thanked him but told him I was going home to help my parents butcher the hog.

He said he would talk to the engineer anyway so that we could get into the locomotive if we changed our minds. I was still not sure if we were being set up or not.

Finally, the train came, and we decided not to go to the engine. Most of us didn't have tickets. When the conductor asked for mine, I pointed to the west. He said, "Thank you." A lot of people did the same. Later on, he came around announcing that if the train stopped on an open track, anyone who had no border clearance had to get off. That would mean that Soviet troops were waiting at the next station.

On the train I was sitting next to a large steel worker. He pinched me on the side and asked me where I was going. I told him I was going home to help my parents butcher the hog. I asked him where he was going, and he said in a loud voice that he was going to Uruguay. Everyone perked up their ears. I asked him why, and he said, because there were 127 religious holidays when they didn't have to work. This broke the tense silence. Almost everyone laughed and started to relax and talk. For the next few hours, it seemed like no one worried if they stepped on a landmine, got captured, or escaped to freedom.

As the train headed toward the border, everybody started to get really nervous. At one point, the conductor came around and announced again that if the train stopped on the open tracks, everyone who didn't have the correct permits should get off. This implied that either the Soviets or the Hungarian Communist border police were waiting to search the train for escapees at the next station.

Nearly everybody started to get ready to depart. At the next station, we decided to take up my railroad worker's offer and get up in the locomotive with some new acquaintances. There were about eight others who begged to come along, but there was very little room in the locomotive. After a while, the engineer stopped the train and warned that we needed to get off. Some of us wanted to give him money or jewelry, but he wouldn't take any.

It was getting dark, and we tried to figure out which way was west without following the train tracks; however, we were unsure. An old farmer was walking by, so we asked him how to get to Austria. He gave us directions through a shallow swamp where the Soviet tanks could not follow. We crossed the swamp and thought we were in Austria. Then a loudspeaker blared out, "This is the Hungarian border guard. You are surrounded. Lay down your weapons and put your hands on your heads."

Laszlo Csele and I were walking together. Both of us instinctively started to run. The last words I heard in my native land were the following: "Stop running, because we shoot to kill." I kept on running anyway and decided to stop only if they shot flares to see me or released attack dogs on me. In the meantime, someone from our group yelled back, "You garbage can. Why would you shoot a fellow Hungarian who has the same blood as you do?" I kept on running, afraid to call out for Laszlo, whom I had lost. Then I got to a cornfield. The corn stalks were cut, tied into bundles, and stacked. I kept going, but I did not know whether I was headed back to Hungary or toward Austria. In the distance I saw a light. I wanted to explore it because it could have been a border police post, Hungarian farm, or Austrian house. The light made it possible to see the backyard, which had no police vehicles or sentries but a cottage and stacks of hay. Then a dog started barking, and I got down and waited for a while.

After the dog stopped barking, I slowly worked my way around the house and listened by the window. An elderly couple was speaking German, and there was a large crucifix hanging on the wall. This was favoring Austria more, because in Hungary, if the Communists found a crucifix that large, it could cost a person his job or worse. I knocked on the window. The man came out, and I asked him where I was in German. He replied, "Österreich, Austria." Thank God I had made it to freedom. My legs wouldn't move. He gently helped me into the house and to the couch while his wife said the *Heilige Maria* or Hail Mary. They offered me food and Schnapps and told me there were not many who made it across the border anymore.

It was November 24, 1956, when I became free. The couple also told me there was a refugee processing center in the nearby village, so I decided to proceed there. He showed me the way until the streetlights of the village. As I started to register, somebody called out my name. It was Laszlo. He had gotten there about half an hour before me, and he was already wheeling and dealing. After a while, he came back and told me, "If we said we were Jewish, a Jewish charity organization would take us to Vienna in a Volkswagen minibus and put us up in a hotel. If we didn't, we would have to go to a tent city camp for months." Then he said we were going in the minibus, and I asked him how, because we were not Jewish. He said that he had told them my mother had been Jewish, so they agreed to take us. I told him I was not going to start my life in freedom with a lie and that we were not going on the minibus.

While we were discussing this, the rest of our group arrived. They had made a deal with the Communist border guards. They gave them all their money and jewelry and were let go. They were very lucky, because we had heard stories that in such deals, sometimes the Communists would take their valuables, let them go, and inform the Soviet troops of their whereabouts. I was very happy to see them, because when we were getting off the train, there was a couple with two small children and a large suitcase. They had a lot of trouble carrying them all, and because I had no luggage, I volunteered to carry their suitcase. During my run, I was afraid to throw it away, because it might have left a track. Beside, if I had been captured, I could have given the suitcase with baby supplies to them. It was a delight to return it to them.

The architect and his wife were also in the group. We all became friends in our escape. Now, Laszlo was trying to convince them to persuade me to take the bus. Their final argument was that all of us could get to Vienna if I went along with the story. I did not have to lie. I just needed to pretend I was exhausted or asleep. This worked for a while. When we got to Vienna, the driver became obviously suspicious of us, because he kept asking me questions about the Jewish holidays my mother observed. Then the car suddenly stopped on a street corner, and they ordered us out. We thanked them anyway.

Chapter IX

First Steps in the Free World

There we were, four in the morning, in magnificent Vienna with brightly lit store windows and cars on the streets. We were very tired. I had practically no sleep for two days. Our clothes and shoes were muddy. We had this idea of asking the doormen in hotels if we could sleep on the basement floor in return for work, but none of the doormen had the authority to allow that.

We were discussing how we could find shelter when a well-dressed couple walked by us and overheard our conversation. They asked us in Hungarian what we were trying to do. When we told them, the man said that he knew a place, and he would take us there in his car. As we were walking toward the car, I suddenly remembered my spy friend who had been caught in Vienna by Communist agents and taken back to Hungary. I told my friends that I would sit behind the driver. If it looked like we were headed toward Hungary, I would grab his neck, and we would jump out when the car slowed down. The car stopped by a hotel, and the driver of the car said to wait until he came back. When he reappeared, he said that the hotel concierge would tell us what to do in the morning. We asked for the kind stranger for his name and address, but he wouldn't give it to us.

When we went inside the hotel, each of us was led to a room. This was around five in the morning. I was so excited I couldn't sleep at first, but when I finally did, I didn't wake up until 11 o'clock. My boots were polished, and there was a razor in the bathroom. I got ready quickly and ran downstairs. When I asked the concierge what work we were supposed to do, he said the man had paid for three-day stays and had also purchased meal tickets in the cafeteria for three days for us. It was like a miracle. After we ate, we went to St. Stephan's Cathedral to thank God for our good fortune. Then we parted from each other.

The architect couple went to the Swedish embassy, and Laszlo and I went to the U.S. embassy to apply for immigration. By that time, over a hundred and fifty thousand people had escaped, and most of them wanted to come to America. The U.S. quota had only been thirty thousand.

After I stood in line for hours, I finally moved inside the door. I was on American soil! In just a couple of days, I changed from a person wanted by the Hungarian Communist police and the Soviets into a free person on American soil. My interviewer spoke good Hungarian, and he knew about my gymnasium in Pannonhalma. When the interview turned to my activities during the uprising, he seemed to know quite a bit about the radio building fight. Then he asked me why I wanted to go to America. I told him, "Now that I am free, I want to live in the freest and best country in the world." I also told him if ever Hungary was liberated, it would most likely be the Americans to help, and I wanted to be a part of that campaign.

When we got to my schooling, I told him I wanted to complete my medical studies and become a doctor. He advised me that if I wanted to be a doctor, I should wait to go to America and finish my studies in Europe and then go to the United States, because getting into medical school there was very difficult and a lot of Americans couldn't get in. In fact, many went elsewhere and returned to the United State when they had become doctors. He also told me that even if I got into medical school, I would probably flunk out because of my very limited command of the English language. I insisted that I wanted to become an American doctor.

Germany and Sweden offered to take Hungarian refugee medical students on scholarship. After my papers were filled out, he told me it would probably be several months before I would get my green card visa. After the interview, I was delighted that at least he had acted like I could come to America, especially because only a minority of the applicants got visas because of the immigration quota.

The day after the interview, I contacted the doctor and businessman whom Professor Petö had recommended. The doctor took me to the Allgemeine Krankenhaus, the world-famous Viennese hospital. The businessman took me for dinner and offered to help me find a part-time job, secure housing, and get me into medical school. I had one more day paid for at the hotel. I started to visit the numerous aid facilities set up in Vienna to help the refugee masses.

In the summer of 1956, when the Communist grip loosened a bit, we started to get tourists not only from Communist-dominated countries but from the West, too. On the black market, my brother and I traded Hungarian *paprikas* for nylon tablecloths and stockings with an Austrian couple. I never expected to see the couple again, but now that I was in Vienna, I recalled their last name.

I went to a Red Cross aid station and asked if I could use one of the phones. There were quite a few names like theirs. The husband's first name was Charlie, but there was no such name in the phone book—it was obviously a nickname. I started to call everybody with the same last name. After about eight or 10 calls, Charlie answered. He cried out, "Laszlo, I am so glad to hear your voice. Where are you?" He told me they were coming to get me, and I was going to stay with them. My guardian angel had smiled on me again, because my days were up in the hotel.

It was a wonderful coincidence. During our conversation, he mentioned several times how he would like to have chicken *paprikas* with *spaetzel*. I was anxious to please them, so I offered to make it. I asked when he would like to have it and he said, "How about tomorrow?" I said okay, and he handed me some money for the ingredients. The only problem was that I did not have the slightest idea how to make chicken *paprikas*.

The next morning I went from aid station to aid station and asked older Hungarian women how to make it and what ingredients I needed. I bought the groceries and memorized how to cook it; however, my other problem was the *spaetzel*. The whole family watched me work. We had a few drinks while I was cooking. Every time I looked at the *spaetzel* dough, I thought it wouldn't be enough, so I kept adding more flour, eggs, and water. I did not know the *spaetzel* swelled in the boiling water.

Finally, everything was full of *spaetzel*, and I still had some dough left. I slipped into the bathroom and flushed it down the toilet. By the morning, the toilet line was blocked by the swollen dough, and a plumber had to be called. When I started cooking, Charlie asked me to let him know when dinner was an hour away from being ready. Before that time, he disappeared. Just as the dinner was ready, the door opened, and he came in with my brother. It was the most joyous of occasions. I did not know where he was. I couldn't find him in Budapest, and I did not know whether he had been killed or captured—or had managed to escape. It turned out that he had escaped 10 days before, and he also contacted Charlie.

I asked him to come with me to America, but he would not, because he there had been a European competition for young opera singers in Austria a couple of days before, and he came in second among tenors. He won a four-year Ford Foundation scholarship to the famous Viennese Music Academy. We made plans to see each other later.

In the meantime, after the U.S. embassy told me that getting my visa would take months, I started to look for a job. My six years of German language study would come in handy. Every day, the U.S. embassy posted the list of people who had gotten their visas. I went by to see if somebody I knew was on the list. After about a week, I saw my name—Laszlo Makk—on the list. It was about six o'clock in the evening by the time I got into the embassy. They confirmed that I was *the* Laszlo Makk who had obtained the visa; however, I was also told that the last bus to the airport had already left, and nobody was going to the airport by car from the embassy, so I could not get a ride. My plane was to leave at 8:30. They then gave me my green card, which would allow me to enter the U.S. section of the airport as soon as possible. They wished me good luck on getting to the airport in time.

I ran home immediately; nobody was there, and I didn't have a key. The housemaster told me that their son was at the skating rink a couple of blocks away. I ran over there, but I couldn't see him. I asked the music operator if he would be kind enough to stop the music and page him. He said they only did that for emergencies. When I told him my reason for asking him, he immediately paged little Johnny, who hurried home with me. When I told him I was going to America, he got quite sad. I left a thank you note for his parents, hugged him, and took the streetcar that went toward the airport. At the end of the streetcar line, I started to hitchhike.

Finally, a truck stopped. He was not going to the airport, but he could take me part of the way; however, when I told him my story, he decided to take me all the way to the airport. I got there around nine o'clock. With my green card, I could get in the American section. My plane was late coming in and had not even arrived yet. There was a big party at the airport for a famous Hungarian actor who was one of the refugees as well. It was given to him by his Austrian colleagues. That was the first time I tasted champagne.

Because our plane had still not arrived, the U.S. Embassy staff provided cots for us to sleep on. Around 11:30, we were awakened and told the plane had arrived to pick us up. We were urged to board quickly. Then the doors closed, and we were on our way. The plane had no signs on the outside, and it was khaki on the inside. The flight was arranged and paid for by International Rescue Agency, and it was part of the airlift carrying Hungarian refugees. I wondered what the International Rescue Agency was, because I never ran across them again. It did not matter. Thank God we were on our way to America.

The next morning, we landed in Shannon, Ireland, to refuel. While the plane was serviced, we were taken to an army mess hall where we had a magnificent breakfast. It was the first time in our lives that we could take anything we wanted on the food line. Orange juice was hardly ever available in Hungary. Here, we could drink all we wanted. We could have all the ham, all the sausage, and all the eggs we wanted.

Most of the servers on the food line were black servicemen. For many of us, it was the first time we had ever seen a black person. They seemed to get a big kick out of the starving mob attacking the food line, and enjoyed piling the food onto our plates.

Eventually, the plane was ready, and we boarded with full stomachs and our first, most pleasant experience with blacks. The engines started, and we were flying toward our dream world, America. To most of us, this was like a beautiful dream rather than a reality.

As we were flying over the Atlantic, a bilingual stewardess called around asking if there was a doctor or nurse on the plane. No one stood up, so I did. I told her I was a medical student. She said that they needed help, because a pregnant woman near the front of the plane had gone into labor. I had assisted with a few deliveries in medical school, and my professors had let me deliver one baby; however, I was now in the middle of the Atlantic with no professors guiding me.

I did not have time to get scared, because the woman needed my help. Because the membrane had ruptured, everything was wet and bloody. We moved her to the front seat, where we had more room, and placed her in a lying position. There were no sheets or towels, only blankets. Then I asked her to push down when the next labor pain came, but she refused, because she didn't want to have the baby until we had landed in America. That way the baby would be American. I had two stewardesses push on her belly as hard as they could to help the delivery, but we didn't make much progress. Thank God we were approaching our next refueling stop at Gander Air Force Base in Newfoundland. The pilot arranged for an army ambulance to meet us on the tarmac. We landed during a snowstorm, and immediately got her down the steps with the help of her husband and the stewardesses. As we negotiated the steps, I hoped the baby wouldn't pop out of her right there and then. We assured her if she had the baby in the ambulance, it would be a U.S. citizen.

In the Gander Airport, we were led to a receiving building where the Salvation Army had doughnuts, hot chocolate (which was a rare delicacy in Hungary), and coffee. Everybody got a little blue bag containing a toothbrush, toothpaste, chewing gum, a Hershey bar, shoe polish, and a brush. We were most grateful for this hospitality so late in the evening. This was the kind way America welcomed us. I still have the little blue bag. Every time I look at it, I say to myself, "God bless America."

We were curious about the wonderful people who wore Salvation Army in golden letters on their caps. We thought they might have been some kind of special army unit. I knew the word army, and I had a good idea what the word salvation meant. We were really concerned that the American army had such old, fragile people, because the Soviet soldiers were much tougher, meaner, and younger. Later on, I found out what the Salvation Army really was, and I never missed a chance to donate money to them after that.

We did not know who our transport plane belonged to, but it did not matter. We were then informed that the next stop would be our final destination—Camp Kilmer, New Jersey. We were in America!

Chapter X

America—First Steps

When the plane landed in Camp Kilmer, an American GI stepped through the door, saluted us, and in broken Hungarian said, "Welcome to America." There was not a dry eye, including the crew. They had brought us to freedom from the hell of a repressive Communist dictatorship.

Shortly after we deplaned, I was paged to come to Barrack 36, where an U.S. Army captain who spoke fluent Hungarian interviewed me. When we finished, he asked if I would like to send a message to my relatives in Hungary. I wanted to send a message home, but I was afraid if the Communists found out where I was, they would punish my parents or sisters for my escape. I quickly composed a message so neither the sender nor the recipient could be recognized. In the family, we used to refer to Pannonhalma as "Holy Hill," so my message read, "To the loving parents at the foothills of Holy Hill, your son is well in America."

My second interviewer was Mr. Andreanszky, an educated American-Hungarian. He quit his job and volunteered as an interpreter to help the Hungarian refugees. He asked me if I would like to help interview incoming refugees about what their job skills were and what they did during the uprising. I said I would be glad to help. The interviewees complimented me on my command of the Hungarian language, because they had thought I was American. I did not correct them.

Occasionally, I found someone who exaggerated his job skills, such as a dishwasher who tried to pass himself off as a chef. One fellow seemed to be a little nervous. Most interviewees were relaxed and happy. He unfolded a Hungarian flag with holes in it and told me that he had taken it from the steeple of the Hungarian parliament. This was strange, because a red flag was flown on the top of that building. It had never been taken off during the uprising, because the steeple was very tall and the top was unreachable without scaffolding. When I asked him when he took the flag, he said, "During the bloody Thursday massacre, and the holes were by shrapnel fragments." Then I asked him what the fight was like, and his description was inaccurate. I knew because I was there at the end of that fight. He was quite talkative and mentioned how the sun was shining on the dead. That did it, because it was raining that day. I excused myself and told my boss about it. He told me to keep interviewing the man until he came in. He showed up with another man, and they took the fake freedom fighter for further questioning and checks. My boss later told me it was determined that he was a Communist infiltrator pretending to be a freedom fighter, and he would be flown back to Hungary. I remarked then that it would be better punishment if he fell out of the plane over the Atlantic Ocean.

On my first day, someone came around asking if anyone was a Boy Scout. I used to be a scout in Troop 627 before the Communist government abolished the scouts. He was asking, because in New York City, there was a Boy Scout convention, and they wanted somebody to talk about the Hungarian revolution. He wanted to take me to New York City that evening to address the convention. Mr. Andreanszky came along to interpret. On the trip, I was fascinated with the beautiful Christmas lights everywhere. We crossed under the mighty Hudson River through a giant tunnel, and then we saw New York City. In my wildest dreams, I never imagined seeing New York City, and here I was in the magical city less than three weeks after I had escaped from the claws of the Communist dictators and Soviet occupiers. The city was unbelievable. I no longer felt fatigue. The magic of the city got me so euphoric, and I could hardly contain myself by the time we turned on to Fifth Avenue. There were lights, shop window decorations, and happy, well-dressed Christmas shoppers all over. I was not even sure that we were on the same planet anymore. Perhaps we had somehow gone to heaven or paradise.

We eventually got to our destination, and I realized I hadn't had time to prepare my speech. I just stood up and told the scouts my personal experiences under the Communist dictators and our fight for freedom. Luckily, I had time to think while the interpreter translated. When I finished, I got a rousing ovation, and many of the scouts offered all kinds of help. I suggested that they help the other Hungarians, because they needed it more than I did.

We got back to Camp Kilmer late, and the next morning, I started my work again. It did not last long, because the Newman's club (a Catholic fraternity) was looking for someone to go on a speaking tour of colleges and civic clubs to raise money and provide scholarships for refugee college students. They selected me. It was a great trip, and we raised a lot of money as a result. After the completion of the tour, the interpreter and handler wanted to take me for a fine meal at a restaurant in New York City called The Russian Tea Room. I refused to go or do anything with the word "Russian" in it.

Then I headed back to Camp Kilmer for Christmas. The Christmas spirit was really high, and it didn't matter that around 36 of us lived in a barracks or that many of us had lost our families forever. We were free! A volunteer took us to a grade school Christmas party. This was the first time I had heard "Jingle Bells." As these kids were dancing and singing happily, I could feel the love and joy of Christmas, and I just started to cry. I concealed it the best I could so that I wouldn't disturb the show. "Jingle Bells" is still my favorite Christmas song.

For Christmas Eve, I was invited to my immediate supervisor's house. It was wonderful to be with Mr. Andreanszky's family for Christmas. I found out that he used to be an army officer, but now he was a carpenter so that he could support his family. I really admired him for volunteering at Camp Kilmer, but his absence at home resulted in some untidiness. Somehow I ended up painting their entrance hall on Christmas Eve.

On Christmas day, I was invited to Mrs. Renee Rasic's house for lunch. She was a refugee from Serbia and wanted to have a Camp Kilmer refugee for Christmas. It was very simple, because she was a woman of little means. She had heard my speeches about the Hungarian uprising and wanted me to have a Christmas visit. She gave me a wool shawl. I still have it, and I'm even afraid to use it, because I might wear it out. Then I went back to Camp Kilmer, where I had a real surprise waiting. The Ford Foundation provided financing for a crash English course for three hundred or so Hungarian refugee university students at Bard College at Annandale during the regular students' Christmas break. I was one of the students selected for this course. Before I left for Bard College, I had a chance to take some sort of exam in Hungarian and was told they would attempt to give me a chance to continue my medical studies; however, I heard nothing further after that, and my hope eventually evaporated.

We were then transferred to Bard College. The campus was beautiful, and we started our language studies with a lot of homework. It was time for me to make a big effort to continue my medical studies. I got the addresses of 10 or so medical schools from the library. I composed a short letter about my background, the courses I had already taken, and my finishing grades. There was a volunteer bilingual priest on campus who helped me translate everything. Then I started to type the letters and course summaries to each medical school, because I didn't want to send carbon copies. Within a week, I received rejections from each school. The priest who had helped me was quite dejected. When he asked me what I was going to do next, I told him that I was going back to the library to look up more medical school addresses and keep applying until I found one that would take me. Little did I know, that I had practically no chance of getting accepted with my undocumented background and poor command of the English language. It did not deter me. After all—I was now in America, the country where dreams became reality.

Then my guardian angel came to my shoulder again. One day, I was walking back to the dormitory after classes when a nice big Buick passed by me. Four of my classmates were inside. The driver was an old Hungarian immigrant lady who volunteered her services to help refugees. When they stopped, I asked them where they were going. They said they were going to some college interview and that they had room for one more. They asked if I would like to come along for the ride. Of course I wanted to. I had never sat in such a luxurious car in my life.

It turned out that we were going to Union University in Schenectady, New York. Dr. Chester Davidson, the provost, interviewed them. The old immigrant lady was trying to interpret, but she hardly spoke any English. We sensed we were getting nowhere. Then Dr. Davidson asked if anyone spoke another language. I told him I knew German. He was a linguist and spoke German fluently. From then on, he asked me questions in German, and I translated them to Hungarian so that the potential students could answer. Everything went fine, and he decided to take all four students into the liberal arts college.

He then asked, "What about you?" I told him my story briefly. He said that I belonged in a medical school. I told him that I hadn't had any luck getting in so far. He asked me if I wanted to go to Albany Medical College. I told him that I would love to. I didn't know at that time where it was and that it was part of Union University. After he made several phone calls, he asked me if I could go there the next day at eleven o'clock for an interview. I told him I would be there and thanked him profusely. The old lady offered to take me. On the way back, we were all overjoyed, because my buddies had just turned into American college students.

The next day, we started out early for Albany because there was a snowstorm. At one point, we slid into a ditch. Eventually, with the help of some good Samaritans, we got the Buick out of the ditch, but we lost time and got to my eleven o'clock appointment at one o'clock. My interview was with Dr. Wolf, the head of the admissions committee and chair of anatomy. Finally, we located his office. His stern secretary told us that we had failed to keep the appointment and that Dr. Wolf was very busy, so we couldn't see him. We kept pleading with her until she finally relented. After a short wait, we were ushered to Dr. Wolf's office.

He first asked for my transcripts. My interpreter lady didn't know what the word transcript meant. Finally, with my Latin background, I figured out "trans" and "scribo," which meant "writing over." We told him that I didn't have any. He told me to get them and then apply for another interview. He started to get up, indicating the interview was over. We started to plead with him, explaining that because of my escape to the West and activities during the uprising, I had no chance of getting any of my transcripts. Also, if the Communist dictators found out where I was, they might force my family to try to get me back to Hungary. Otherwise they would be severely punished.

Then he took my homemade transcripts and seemed pleased with the grades. Next, he started to ask me anatomy and histology questions. Again, we had trouble with translation. I asked him if I could say the answers in Latin, because I had had eight years of Latin and had learned all medical terms in Latin in Hungary. For example, spinal cord was *corda spinalis*. Now he was smiling, and after several microscopic slide reviews, he said, "Welcome to Albany Medical College. You can join the junior year in September, but learn as much English as you can between now and then." On the way back to Bard College, I was on cloud nine. After all the adversities, I had one more chance to become a doctor, but not just any kind of doctor—an *American* doctor.

Then another problem surfaced. I realized after a while that I had nothing in writing about my admission. Suppose Dr. Wolf got hit by a bus or had a heart attack. What would happen to my admission then? I decided to move to Albany, try to get a job, and keep going by the medical school every chance I got to remind them that I was coming in September. The refugee placement center at Bard College found somebody in Albany who would take me until I found a job. His name was Mr. Matthew Bender, the senior executive at the Matthew Bender Law Book Publishing Company.

On a Sunday morning in January, I took the train to Albany. I was given the phone number of his nephew, Mr. John Bender, to call when I arrived, and he would pick me up at the train station to take me to Mr. Matthew Bender. There was a snowstorm, and the train station was very cold; however, I didn't feel it, because I was too excited about my new home. Mr. and Mrs. John Bender picked me up and told me that I was not going to "Uncle Matt's" house. Instead I would go to their house. Later on, I found out that Uncle Matt backed out of taking me in after his cook told him that she would quit if he brought any foreigners into the house.

John and Faye Bender had a magnificent mansion in Loudonville, a suburb of Albany, seated on 35 acres. They had four children, and they had already taken in more than 10 Hungarian refugees. They were warm and caring people. I kept a warm relationship with them for more than fifty years. After a while, they introduced me as their foster son and Carolyn as their daughter-in-law.

This paradise at the Benders' loving household started to show signs of some problems. My fellow refugees were a mixed group, ranging from freedom fighters to lazy bums who had not participated in the uprising and who had left Hungary when the border had been opened to seek a better life. One of us, Paul Györi, was in a Communist prison camp for 10 years. His fingernails had been pulled off, and his teeth had been knocked out during the torture sessions. He was a great patriot, and the Communists never broke him.

During the uprising, when the prisons were opened, he became free and escaped. He was a talented machinist and a bad alcoholic. The Benders found him a job in the Colt gun factory in New Haven, Connecticut. There, he invented and patented an improvement on Colt handguns. He was well compensated, and he received generous royalties for his invention. He became such a bad alcoholic that the Colt plant told him not to come near the plant anymore, that they would mail him his paycheck and royalties. He eventually drank himself to death.

There were others in the group who did well and got entry-level jobs after they learned enough English. One was Marty Nagy. He started out as a janitor in a bank, and he eventually became vice president.

At the beginning, there were some adaptation problems with the Benders. Even though we had several bathrooms at our disposal, some of the men would not bathe, shave, or even use deodorant. One night after supper, Mr. Bender brought us all in the library and closed the door. There was a row of deodorants on the table. Mr. Bender then explained that each of us should take one and use it every day, and he then showed us how. After I explained to Paul that we needed to stick this in our armpits and move it around, Paul told me to tell Mr. Bender that he would rather stick it up his '*bleep*'. Mr. Bender had been a major in the U.S. Air Force working in intelligence. Because he was a good face reader, he suspected defiance. He asked me to translate what Paul had said. I told him that I needed to explain the deodorant to Paul a little better first. I then told Paul that he had no choice, that it was a house rule, and he either had to start using it or start packing, because he would be kicked out if he didn't. Finally, he got the message and obediently started to use the deodorant right there.

I became quite concerned that we were becoming a burden on the Benders' goodwill. Therefore, I started to look for a job with the Benders' help. The first job opportunity I received was taking care of Mr. Page, who was a paraplegic. I did not get that job, but then the Benders found an opening to for a janitor and aid combination at the Cerebral Palsy Treatment Center for Children. It was an ideal job for me, because it was on the medical center campus next to the medical school. Fortunately, I did get that job, which involved cleaning the floors and treatment tables, taking the children to the toilet, taking their braces off and putting them back on, taking them to their cars when they were picked up in the afternoon, and any other gofer job my limited English allowed. Despite that, I loved it. The director apologetically explained that they could only pay $27 a week, and I informed her that I was very happy with the pay, because it was much more than my professor of surgery in Hungary made.

The building where I worked was on the medical center campus near the medical school, and I used every opportunity to walk through the school to remind the dean of students that I was coming in September. My transportation was easily arranged, because either the Benders or their chauffer, Arthur, would take me or pick me up in a Lincoln or Mercury. The Benders, with their big hearts, offered to let me stay with them while I went to medical school. I was most grateful for their offer, but I decided to decline. I didn't want to be an imposition. In fact, I felt them driving me to and from work was an imposition. Consequently, with their help, I started to look for a place to live where maybe I could work off my rent. The Benders found a place at Miss Harris's house. She was the headmistress for the Albany Academy for Girls, a 200-year-old prep school. She had a 90-

year-old invalid mother, and my job would be similar to that of a houseboy. It basically amounted to a maid and practical nurse. I was very happy about the job initially because I essentially had a second job that provided food and a place to sleep which allowed to keep nearly all of my first job's salary. It was also conveniently located close to the medical school.

The Benders took me over to Miss Harris's house on February 13, 1957 at around five o'clock. It was snowing heavily. After I got inside the door and put my belongings down, Miss Harris gave me a shovel to clear the snow. By the time I had finished at seven o'clock, my hands were cold, and I was really hungry. I was looking forward to getting inside, but that did not happen. When I knocked on the door to come in, Miss Harris came out and told me to clear the snow off the driveway to the garage, which was at the end of the backyard. I finished at nine o'clock. When I got in, Miss Harris showed me to my room, which was a cold porch on the second floor. It was so narrow that when I opened my arms, I could touch the wall and the window. I was cold and really hungry, but I was happy as well, because I was free and self-supporting. Furthermore, I did not have to worry about the secret service knocking on my door.

In the morning at breakfast, she told me to come home on my lunch break to feed her invalid mother and the dog. I also had to let him outside and make sure he went to the bathroom and then let the dog back in. After that, I could make myself a sandwich. My place of work was a 15-minute walk each way, which made it difficult to cram all of these chores into a one-hour lunch break. Particularly, when the dog did not want to go outside and do his thing. The longer the dog waited, the faster I had to walk back to work, which could be quite a challenge in ice, snow, or rain.

After a while, we settled into a routine. I helped with dinner, served it, and cleaned the table and the dishes. Before dinner, I brought "grandma" downstairs. After dinner, I took her over to the living room to watch television, and at nine o'clock, I took her back upstairs.

After Miss Harris learned to trust me, she availed herself to accept dinner invitations from her students' parents several times a week, because now she had a free sitter for her mother. It was not the most exciting job, but I had great benefits from it. The old lady talked slowly in very correct English, so I could learn the language from her. In addition, I helped with the laundry, yard work, and anything Miss Harris could think of.

On Sunday afternoons, I used to go to the International House to meet other students from various parts of the world. When spring came, other Hungarian refugees who had cars used to pick me up, and together, we would go for picnics. However, Miss Harris soon stopped that. She told me I could only go out once a month, because I would get too tired if I went out more than that. Every Sunday from then on, she always had a job for me inside—polishing silver, doing laundry, or washing the windows. After a while, some of the neighbors asked what she paid me for all that work. When I told them, they just shook their heads. I was not too concerned with it, because I was closer to my goal in life, becoming a doctor.

Eventually, I opened a savings account at the bank, and I deposited all my paychecks, though I kept a few dollars for basic necessities. I figured if I didn't have it, I couldn't spend it. Occasionally, when I was really hungry, I treated myself to a pack of hotdogs or bologna, which I finished eating on the way home.

I kept up my friendship with the Benders, and I was always free to go to them for dinner or a movie. They were absolutely wonderful to me. In the meantime, one of the speech therapists got me another job tutoring German on the way home twice a week. This was great, because while I was teaching my pupil German, I also learned English from him, and he was paying for it, too. Some of my fellow Hungarian refugees ridiculed me for working so hard. They thought I just didn't want to spend time with them. But I just knew where I was going and made every effort to get there. As time passed, I got in touch with my brother in Austria. I was very homesick at Miss Harris's, because at the age of 24, I was practically chained to the house for no pay.

Christmas was very festive. Just as I started to wash the dishes after dinner, the phone rang. The caller was Mr. Bender. He was a practical joker, and said that he had forgotten to get cash from the bank and asked if I had $16 to loan him. He said he would come by to pick up the money and me. I did not have that much cash, but I still had 10 minutes before the bank closed. I told him, "Of course, I have the money for you."

When we hung up, I ran to the bank to take 16 dollars out of my savings. When Mr. Bender came to pick up the money, I was free to go, because I had already finished the dishes. When we got to the Benders' house, the entire family was there, including Mrs. Bender's aunts from Texas. John Bender asked for the 16 dollars. When I gave it to him, he gave me a car key and said, "Merry Christmas."

I couldn't believe it. There was a two-year-old Buick parked outside. He said that it was now my car and that the 16 dollars was for the new license plates. I could not drive it yet. I was up all night that evening thinking about accepting this generous gift. I came to the conclusion that

I could not afford the gasoline or the extra expense of keeping a car. I also figured that I could easily walk to school and work. If I had a car, it might take me away from my studies when I was struggling to avoid flunking out. With all of that in mind, I decided not to accept the car.

On Christmas morning, I told Miss Harris about the great surprise. She told me I didn't deserve the car, and that indeed I shouldn't take it. Nobody had ever given her anything like that after all. I wished her a merry Christmas anyway. The Benders were a little disappointed when I told them I couldn't take the car. Mr. Bender told me they were moving to New York City, and they were only taking two of their three cars. This third car was Grandpa Bender's. He asked me once again if I would take the car. I thanked them and told them I could not. It wasn't until I finally became a doctor and bought my first car that I told them the real reasons why I had not accepted their gift.

Life moved along as I continued to pursue my medical degree. One day I received a letter from one of my old classmates in Budapest. He informed me that if I returned to Hungary, I would receive complete amnesty and would not even lose one year's worth of credit of medical school studies. He also wrote about how the regime had changed to a more democratic one and that Hungary needed smart people to build it back up. I actually was very homesick. At times, I didn't sleep at all. I just listened to the rain on my porch's metal roof. My dream was to return to Hungary as a member of the U.S. Army to liberate my homeland. This offer was very tempting. I could become a doctor back at home in Hungary, help my fellow countrymen, help my parents in their old age, and do the best for my country as a doctor. I decided to go back, but I did not want Miss Harris to know my plan. I would just disappear—a payback for all my frustrations.

The next weekend, I took a train to Washington, went to the Hungarian embassy, and requested a visa. They gave it to me promptly. After I returned, I went to see Mr. Berman, head of the immigration service in Albany. I told him what I wanted to do and why, and expressed my gratitude for America's great hospitality. He said if that was what I wanted to do, they would fly me back. I had saved four hundred dollars, and I wanted to pay for the trip myself. That was the least I could do to express my gratitude to America. Then Mr. Berman made a couple of phone calls and told me to wait. The next thing I knew, the Benders were there to pick me up. Mr. Bender told me that it was a Communist trick for sure. Because he had been in the intelligence service during WWII, I trusted him to know this.

I gave up my Hungarian visa and decided to stay in the United States at that point.

It hit me hard to think I came so close to losing my freedom as well as my chance of becoming a doctor—both of which I cherished so much. I shared this story with my family in Hungary. Years later, my sister, she ran into my former classmate who had supposedly written the letter about my amnesty offer. He told her that he had never written any such letter and hadn't even known where I was. Evidently, this was indeed a Communist trick to get me back to Hungary.

To speed up my pace in learning English, I cut pieces of paper into small squares and wrote the words in English on one side and in Hungarian on the other. I carried these little papers in one pocket and kept looking at them everywhere I could. Once I knew the word, I put the slips in the other pocket.

In the summer, the regular janitor, Ralph, left the Cerebral Palsy Treatment Center to work on his farm. This was great for me, because I got to wash and polish all the floors on Saturday, and I got paid overtime for my extra work. When September came, it was time to say goodbye to the little handicapped children and staff at the treatment center and start medical school.

Before I started school, I used to go by the medical library to learn everything I could about American medical schools. I found out that pharmacology was in the sophomore year. In Hungary, I studied pharmacology in the decimal system (not the U.S. system). I went to the Dean of Students and asked him if I could start in the second year instead of my junior year. I was worried that not knowing the correct doses in prescribing medications might result in me hurting rather than helping my patients. He agreed, and said if I also take and pass the National Board Exams, that I would get a regular MD degree at graduation. I thanked him, but did not have the slightest idea what National Board Exams were. I was soon to find out.

Chapter XI

Albany Medical College

There were some joyous occasions in my life. One was when a letter from Cornell University Medical School notified me that, based on an exam I took at Camp Kilmer, I had been accepted to continue my medical studies there. This was a great honor, but I had already been accepted to Albany Medical College. Albany was so good to me that I decided to stay in Albany and turn down Cornell even though it was more famous. With some help, I wrote a nice letter to Cornell thanking them. Turning Cornell down may have saved my chance to become a doctor, because I almost flunked out of Albany, and I would have likely flunked out of Cornell, as other Hungarian medical students had.

Another joyous occasion was my visit to Niagara Falls. One of the physical therapists was from Buffalo and was getting engaged to a medical student. They took me along so I could see Niagara Falls. It was almost unbelievable that I was going to Niagara Falls. In Communist Hungary, nobody could go near the border. Here I was walking across the bridge to and from Canada all day. On the way back to Buffalo, I got lost and almost ruined the engagement party.

After so many setbacks and roadblocks, I was going to be a medical student again. It was such a wonderful feeling to be not only a medical student but also an *American* medical student. American medical schools have always been the best in the world. I could hardly wait to start. All my classmates were extremely nice. My conversational English was fair, and I thought I would just be studying the subjects I had already taken in Hungary but in English. I could not understand most of the lectures, with all their technical terms and fast tempo. In the textbooks, the sentences were full of words I didn't know. I studied one sentence at a time. First, I looked up the words I didn't know in the dictionary, and then I memorized them. I studied the sentence, and once I knew it, I went on to the next. This was a daily reading and learning process covering fifteen to twenty pages. When I got really sleepy, I studied standing up so that I wouldn't fall asleep. When I couldn't stand anymore, I often took a cold shower. All of this was to no avail, because I flunked all my first quizzes.

At the medical school, we had pigeonholes for our mail. I never got any mail, because there was no one who could write me, but I checked for mail like everybody else anyway. Finally, I got a letter from the registrar. Basically, it said my academic performance was unsatisfactory and to please improve it. Otherwise continuing my studies may not be possible. A few weeks later, another letter from the registrar came that said I had not shown any improvement in my grades and they urged me to make every effort to improve them. Unless I showed significant improvement in my studies, I would have to be terminated. There I was again. My last chance to be a doctor would be blown away by flunking out.

On top of this, there was a note from the bursar to come and see him one day. I did not know for sure what a bursar was exactly, but my classmates explained it to me eventually. I went to see him, and he got to the point right away. He asked how I proposed to pay for my tuition. I had never paid tuition before, so I asked him how much it was. He told me it was over $1000.00. I almost fainted. I told him I had $300.00 left to live on after I bought my textbooks, purchased lab coats, and rented a microscope. He went through my finances, my jobs, and paychecks. Then he told me he would be in touch. It was a nightmare, the monsters of flunking grades and tuition coming at me. This was before student loans were in vogue. Again, the Benders very graciously offered me a place to live as well as to pay for my schooling. However, I thought I would try to do it on my own.

A while later, the bursar called me in again and said I would have a free tuition scholarship. I just needed to work on my grades. Hallelujah! Then I passed the first quiz in microbiology. Eventually, I started to pass all my quizzes—that is, except for pathology. We had essay exams in pathology. Here, I not only had to understand the questions, but I had to write the correct answers with the correct spelling. Finally, Christmas and the semester closing pathology exam came. All my classmates thought it was easy, and so did Dr. Wright, the department chairman. He came to the lab with the corrected exams just before we started the Christmas break. He wished us a merry Christmas and said these were some of the best exam scores in his 35 years of teaching. He said the worst grade was a B, minus one person who flunked again and whom he didn't need to name—everyone knew who he was. I was the one, but why did he have to rub salt in my wounds? The good news was that I nearly flunked out but didn't.

The Benders were going to spend Christmas at Lake Placid, the famous resort, and they invited me to go with them. Because it was the Bender's invitation, Miss Harris said I could go, which was great to hear. The Benders' car was full, so I went with Mrs. Bender, Sr., in her chauffer-driven Lincoln. I had a gray flannel suit and a pair of khakis. We were supposed to go skiing; however, nobody wore khakis, and they only rented skies, not pants. I decided to wear my grey flannel suit pants. I planned to be really careful not to get them wet or tear them. I did not tear the pants, but they were soaking wet.

I needed to dry them fast so that I could wear them for dinner. I placed them between two towels under my bed sheet and then laid down on them, attempting to dry and "press" them with my body weight. The Benders asked me to come over to their suite for a drink. I had to leave my wool pants on the bed between the towels in order to dry more. While we had a drink, the maid came to my room and made up my bed. I almost cried when I found my wet pants all wadded up looking worse than when I had started. I had no choice but to wear them. Even if they had noticed the pants, no one made any remarks about them. I had a wonderful dinner and danced afterward. I quickly realized nobody was looking at my wet pants.

Mr. Bender had to interrupt his vacation for a couple of days to go back to Albany on business. He flew back on a chartered plane that landed on the frozen ice of Lake Placid. Because I had to get back to medical school before the Bender's vacation was over, they put me on the plane on its return trip to Albany. If there had ever been a dream vacation, that was it—a chauffer-driven Lincoln, Lake Placid with its natural beauty, and taking off from the frozen lake on a chartered plane. All of this, barely thirteen months after I had arrived in the United States as a penniless immigrant. Furthermore, I survived the first semester and didn't even flunk out of medical school.

The second semester of my sophomore year was coming up. The big black cloud over my head was pathology finals. Furthermore, the first part of the National Board Examinations was coming up at the end of the semester, if I survived that long. I knew if I didn't pass pathology, my lifelong goal of becoming a doctor would be over.

I kept studying and studying. In one of the next pathology exams, one of the questions was the following: "A 17-year-old Negro girl has a lump in her breast. What is the likely diagnosis, and how would you treat it?"

I didn't know what the word "lump" meant, so I raised my hand and asked Dr. Wright, the pathology professor, what it meant. He told me that I should know that. Then I asked him if I could look it up in my dictionary. He said that I couldn't because no aids were allowed during the exam. I kept wondering what the word lump meant. Then I remembered that we had had lectures on breast diseases and syphilis since the last exam. I could hardly understand any of the lectures, but I remembered that the words "Negro" and "syphilis" had been mentioned together. I thought I had the answer: lump probably meant a form of syphilis in the breast. I wrote the answer as best I could—the likely diagnosis was syphilis, and I would treat it with penicillin. I even recommended that the doctor should repeat the syphilis test in six weeks if it was initially negative, because it took that long for the test to register positive after exposure sometimes.

A few days later, Dr. Wright's secretary came to the lab with grave concern on her face and told me to report to Dr. Wright immediately. I went right away. When I went into the office, he immediately started to yell at me. He said, "Makk, where did you go to college?" I started to answer that I had not gone to college in this country, but I only got out, "I did not go to college," when he cut me off. I could not say anymore.

He started to scream at me, "You snuck in here without a college education, which I should have suspected from your poor spelling." Then he moved on to my answer on the exam question concerning the lump and started to scream at me that I wasn't morally fit to be a doctor based on my answer. I could not understand most of this screaming, but I did understand clearly his last sentence: "Makk, I don't care who got you in here. I am the one who is flunking you out." After we got our exams back, my classmates explained what a lump was and why Dr. Wright was so upset. The story was so funny that it resonated through to class. Now I got a lot of friendly smiles.

My nightmare with Dr. Wright was not over yet. My landlady, Miss Harris, was a 62-year-old spinster, and Dr. Wright was a 67-year-old widower. Miss Harris kept asking me how Dr. Wright was. She did not ask about any of my other professors. After her tender interest, I couldn't tell her that he was my nemesis. I just told her that he was okay. Then, late one night, some of my classmates and I were discussing whether or not it was worth me studying day and night, because Dr. Wright was probably going flunk me anyway. Toni Arena, one of my classmates, said, "Dr. Wright is a son of a bitch, but he will pass you." I didn't know what a "son of a bitch" meant, but I thought that maybe it was a title, such as duke of Marlboro, and that maybe his nobility would be gracious and not flunk me. The next time Miss Harris asked me about Dr. Wright, I was so glad to have some news. I asked her if she knew that Dr. Wright was a son of a bitch. Her face turned red just like Dr. Wright's, and with the sternest words, she told me not to ever use that expression again. She said if I used it again, I would be dismissed. When I explained this to my classmates, they roared with laughter.

I had so many guardian angels. It seemed like one appeared every time I was in crisis. My new guardian angel was Dr. Arthur Stein, professor of surgical pathology. One day, he told me he knew what I was going through with my language barrier. He was once a visiting professor in South America, and he could not speak Spanish at the time. He understood how I had had trouble communicating information. He said he was going to get me some old pathology exams to practice with, and afterward I could review the correct answers. He said he thought I had the knowledge but just didn't know how to take written exams, especially because all exams were oral in Hungary.

After I got the hang of the exams, another ray of hope appeared. I was informed that Dr. Wright agreed that the grade of my final exam would be my grade in pathology. If I flunked, I was out; however, if I passed, I could continue my studies. That was real pressure. My whole medical career depended on whether or not I flunked yet again. Finally, we got our grades back, and I passed with a C. Hallelujah! I also passed the finals of the other subjects.

Now I had to concentrate on the National Medical Board Exams. These were not mandatory exams at that time. They consisted of three parts. Part I included the subjects covered in the first two years. Part II addressed the third and fourth years of medical school. Part III was given at the completion of an internship. Albany Medical College, along with other good schools, strongly recommended taking these exams. They were comparative measures of national standing in medical education. Feeling it was important, I went on and signed up to take the board exams.

The board exams were approaching, and I had yet to have a chance to even open the school's books covering freshman year material, because I had put all my effort into not flunking out of the second year. I took these exams in competition with students who spoke English fluently and had recently studied the freshman subjects that I had not seen in years since taking freshman classes in Hungary.

One day, I had a note in my pigeonhole to report to the office of Dr. Nelson, the dean of students. Dr. Nelson told me that the National Board Exams would be hard, and many American students even flunked them. He said I would probably flunk every subject, but that I should not give up hope, because I could retake them any number of times. That, again, was a ray of hope. I sat for the exams, and indeed, they were hard. Finally, the results came back. It was a great day in my life. I didn't flunk all of the subjects. I passed four and flunked only three. I felt very smart, because one of my American classmates had flunked four, which was one more than I had. From then on, I never flunked another test in my life.

I was elated, but now a new challenge was coming over the horizon. I did not flunk out, but I didn't have money or a summer job. I was getting desperate, so I went around knocking on doors and asking if people needed their basements cleaned or if they had any other jobs. People were kind, but they did not have anything. Somebody told me that they needed a yardman at Rensselaer Polytechnic Institute (RPI), so I took the bus there immediately. They were very nice, but they turned me down. Then someone suggested I go down to the New York State Health Laboratories, because they might have needed some helpers. After I filled out an application, I got an interview. One of the questions I was asked concerned what I knew about electron microscopy. At the time, I didn't have the slightest idea what it was. I told them I would like to know more about it.

Then I was sent to see Dr. Edwards, the director of the electron microscopy laboratory. He explained they needed someone to tediously clean the glassware using acid and detergents. I got the job, which amounted to a glorified dishwasher. I was delighted. I kept doing everything the best I could, and pretty soon Dr. Edwards started to use me for other tasks. I was not only cleaning the glassware well but with great care. Eventually, Dr. Edwards promoted me to research assistant, which granted me a raise. After school started, I could come in and do my work on the weekends or the nights. This was absolutely great—being a research assistant in a very prestigious laboratory when a month before I was desperate and unemployed. America really is the land of opportunity.

My first assignment was to search for an electron microscopy staining technique that could stain the neuromuscular junctions, where the nerve endings give the command to muscle fibers to move. We knew there was a substance called acetylcholine, which carried this function out. We needed a heavy metal with a similar molecule that would produce a good electron microscopic image. Lead, being a heavy metal, gave an excellent contrast by blocking the electron beams in electron microscopy. Thus, we searched for a lead containing molecules similar to acetylcholine.

After I searched the literature for quite a while, I found that tetraethyl lead, the no-knock chemical used in gasoline, had a very similar structure to acetylcholine and, of course, had lead in it. Our next task was to figure out a way to deliver this substance to the neuromuscular junction and select various solvents for tetraethyl lead. Dr. Edwards knew that the largest neuromuscular junction was located in pigeons. Next, we had to get various approvals from intra- and extra-institutional authorities to use tetraethyl lead. It was quite a challenge, because tetraethyl lead was very toxic.

Then it was my job to inject our solutions into the wing artery of a pigeon and humanely sacrifice it afterward. There were a number of solvents that did not work, but I finally hit a homerun after testing. We got the most beautiful and detailed electron microscopic pictures of the neuromuscular junctions. Now we had to find a way to minimize or eliminate the extreme toxicity of tetraethyl lead. In the meantime, somebody published a lead staining technique that was not toxic, so our lead staining endeavor joined the thousands of other experiments that almost succeeded, and the project was dropped. I did, however, obtain invaluable experience in research.

In September, I started my third year in medical school. I also became a part-time phlebotomist, drawing blood and starting IVs. It turned into a great job, because nobody wanted to work Saturdays and Sundays, and I had no class on weekends. Often I could work two shifts from 6:00 a.m. to 10:00 p.m. It was great, because the pay was double for the second shift and it usually wasn't that busy, so I could study, too.

School was great, because I no longer flunked tests, and we were now involved in patient care. Because I had two jobs, which were very good, it was time to say goodbye to Miss Harris and find a better living arrangement. I rented a small room near the medical school in a private home, The owners, Arthur and Ann Brooks, were simply wonderful. They had converted their attic into three small rooms for the purpose of renting them out. Both of my roommates were also medical school classmates and very smart and helpful. I kept my research assistant job with Dr. Edwards but didn't work as much.

One Sunday morning, I was working in Dr. Edwards's lab, and I needed a chemical from the walk-in freezer. I failed to secure the door, because I thought I would just run in and then run out before the door closed. Unfortunately, it shut before I got to it. I was now locked in the freezer, and there was no one to call for help and no way to open it from the inside. Suddenly, the prospect of freezing to death dawned on me. I kept moving and kicking the door, but the chances of me being discovered in the empty lab were slim. Luckily, a security guard was walking by and heard me kicking the door and let me out. It was amazing that after I had survived a world war, an uprising, and an escape—my life almost ended in a walk-in freezer.

In medical school, I loved every medical specialty I rotated through, particularly obstetrics. Dr. Edwards started to study the electron microscopic morphology of cancer cells. He was particularly interested in lung cancer composed of the malignant variety of squamous cells that were like the cells of the skin. He wanted to get hold of human squamous cell tissue that had never been exposed to anything like the sun. I suggested that I might have been able to get tiny tissue samples of newborn foreskin. At that time, most newborn boys were circumcised right after birth. I had my little vials of phosphotungstic acid electron microscopic tissue fixative in a wooden box, and every time I was around when a baby boy was circumcised right after birth, I asked if I could have a tiny fragment of the foreskin. I am sure a lot of people in the delivery room found it weird that the strange guy with the strong foreign accent was begging for a sliver of newborn foreskin. The electron microscopic pictures were spectacular, and provided us a great way to compare them to lung cancer. We published these findings with Dr. Edwards in the *American Journal of Pathology*. This was only two years after I had almost flunked out because of my poor English.

My interest in obstetrics kept getting stronger. When Christmas break came in my junior year, I had no place to go, so I volunteered to help out in the delivery room to gain more experience. Dr. Robert Nesbitt was the 32-year-old chairman of the obstetrics department back then, and he was a hotshot from John's Hopkins University, known to put down not only medical students but residents and attending obstetricians, too. When I volunteered to spend my Christmas in obstetrics, someone asked me if I would measure and weigh all placentas for one of Dr. Nesbitt's research projects. After I agreed, I was told each had to be weighed and measured three times, and then the results were to be documented.

On Christmas day, we had 14 or 15 deliveries. Neither the resident on call nor I had time for breakfast or lunch, let alone a cup of coffee. In the middle of the afternoon, Dr. Nesbitt showed up and found a placenta that was delivered twenty minutes earlier and had not been measured yet.

He called me everything from stupid to lazy. When I told him that I had to run to the blood bank twice for blood and had to administer it in the last 20 minutes, he said that was still no excuse. When he finished; I had a strong desire to tell him Merry Christmas anyway, as well as punch him in the nose, but I just stood there. When the resident tried to tell him how helpful I had been, Dr. Nesbitt was not interested in hearing it. This poured cold water on my enthusiasms for obstetrics.

My next interest was heart and chest surgery. It started when a child swallowed an open diaper pin. One of the thoracic surgeons let me help to remove it. After the child was put to sleep, he passed a tube into the esophagus next to the pin. He told me to take over and push the pin down to the stomach with an instrument, move it around until it turned so the open end was down, and then pull it out slowly. I got the safety pin out. It was magic, and from then on, I was going to be a thoracic surgeon.

In the summer between my junior and senior year, I got a job with the thoracic surgeons as a gofer. I got to help with the heart-lung machine, cleaning, and set-ups. I got to see or help with most surgical procedures. It was wonderful. One of my professors, Dr. Ralph Alley, was a superb teacher. With him, I started rounds at the radiology department to see the X-rays on his patients. Then we went to see the patients and later review removed diseased organs and microscopic slides prepared from them in the pathology laboratory. It was a great summer of learning, and I earned some money, too.

I had my first American date as well. There was a movie called *Dracula* playing. I did not know who Dracula was, but I knew the movie was filmed in Transylvania, which used to be a very picturesque part of Hungary. I thought the movie was about Hungary. There was a student nurse who was nice to me, so I eventually worked up the courage to ask her if she would like to see a very nice movie on Saturday. She said yes! I planned to pick her up at the student nurses' dorm and take the bus downtown. It was raining pretty hard that night. The first thing she asked was where I had parked. I told her that I didn't have a car but that I had a good umbrella, and we could walk to the bus stop and take one downtown. I could tell right away that she was not thrilled with the idea. We made it to the movie theater, and we only got a little wet.

The movie had beautiful Transylvania scenery. I started to whisper to her that this was my country and pointed to myself to help her understand it. Then Dracula started to impale people. She jumped up and took off. I could never find her, and I did not know what she actually understood from my strong Central European accent; however, that was definitely the end of my first date.

In the fall, I started my senior year. Because of clinical rotations, I could no longer work in Dr. Edwards's lab, but we stayed good friends. He thought I should not try to become a heart surgeon. Instead he thought I should get a PhD and become a researcher. He offered to get me into one of the finest PhD programs in Boston. I decided to aim for surgery, but I still assisted him in reading the galley proofs of the first electron microscopic atlas that he edited.

In my senior year, we had four months of elective rotation, which we could have extramurally. I heard from the Benders what a great cancer center Sloan Kettering Cancer Institute was. It was where Mr. Bender had undergone two cancer operations. The institute had a four-month program for externs. This was a very fine program, and it was extremely hard to get into. I applied anyway and was selected.

During these four months, I had a chance to learn from the most outstanding cancer experts of that time. It was very exciting, and I usually stayed in the library until it closed in order to read up on the day's cases. Another nice thing about this program was my quarters. They did not have room at the intern quarters, so they put me up in the Sutton Hotel. It was not a fancy place, but it was a hotel in New York City.

I was now getting trained in one of the best cancer center in the world. People from other countries dreamed about New York City, and I was living there in a hotel. I worked with some of the greatest living cancer experts, people about whom I used to read in medical books with admiration. Now I knew them and worked with them. My appetite for cancer knowledge was insatiable. When I was not working, I was in the library studying up on patients. This was just like heaven.

My personality was enhanced by my hyperactive thyroid gland for nearly a year, which did not respond to medication. My doctor was the professor of endocrinology at Albany Medical College, and he was determined to treat me medically so that I could avoid surgery. In the meantime, my goiter kept enlarging. My neck size went from a 15 1/2 to a 17 ½. The size 17 ½-inch shirts were so large elsewhere on my body that the shoulders went down to my elbows. At Sloan Kettering, one of the famous head and neck surgeons noticed my large neck and asked what I was doing about it. He said medical treatment would never work, and I should have it taken out. I planned for a thyroid operation after I returned to Albany.

Chapter XII

Planning for the Future

As time moved on, I had to start thinking about my internship, which is one year of clinical training after one receives a medical degree. The internship year is crucial in the development of a young doctor and has a great bearing on one's future. For me as for others, it was very difficult to get into the best places as there were many applicants for few openings. And usually, the better the place was, the less the salary or stipend.

There was a national intern-matching program where the hospitals listed their picks of interns in the order of their preference. If a place considered you a potential candidate, they granted you an interview. They were not supposed to tell you if they picked you, and they were not supposed to ask where stood.

I selected some of the hospitals I knew were excellent teaching facilities. One of the places I applied to was the Cornell Surgical Division of the world famous Bellevue Hospital in New York City. I knew that it was a very competitive position and that there were dozens of highly qualified applicants for each spot. To my surprise, I got an interview. Several professors interviewed me, the last of which was Dr. Cranston Holman, the head of the surgical division. At the end of the interview, he said that they would take me if I listed them as my first choice. I was shocked that I could be accepted to this excellent and prestigious institution, but I also was shocked that he had asked that I list them as my first choice, because they were not supposed to ask that. Of course, I assured him that I would be listing them as my first choice. I was in seventh heaven, thinking that I was going to get my first choice, which was a dream internship, and I didn't even have to sweat the matching results.

After my externship at Sloan-Kettering was over, I returned to Albany to finish my senior year. I also got part-time evening job in the medical school library from 6:00 to 10:00 p.m. when I was not on call. This was a great job, and when things were quiet, I could study quite a bit. When I checked in the returned periodicals, I used to browse through them. They were full of new medical information. One thing that caught my attention was Dr. Michael DeBakey and Dr. Denton Cooley's fantastic heart surgery publications from the Texas Medical Center. It looked like they were the best anywhere. I didn't think I had any chance to get into their heart surgery program, but I applied anyway. When I mentioned to people that I had applied, they said I was dreaming. Even my professor of surgery advised me gently to make sure I had picked some good second and third choices. Nearly everybody thought I wouldn't even get an interview. To my great surprise, I got an interview, but there were two problems. The interview date was a week after my scheduled thyroid operation, and the other was that I didn't have the money to make the trip to Texas.

The Benders were like guardian angels to me again. When I woke up from my thyroid surgery, they were sitting by my bedside. They had come up from New York City. John Bender said that they were going to Texas to visit Mrs. Bender's relatives, and if I would help them with the drive, they would buy me an airplane ticket to fly back. We left just a few days after my surgery so that I could do the interview during my recovery period. Driving to Texas was absolutely great in a Lincoln. Of course, they wouldn't let me drive a single time. I think they didn't want to risk their lives with my driving. After a while, I realized the Benders, with their love and generosity, never intended to let me drive. They just wanted to give me a chance to get to my internship interview in Houston at the Texas Medical Center.

My first trip to the South brought back old memories. In 1945 in war-torn Hungary, the book *Gone with the Wind* (translated into Hungarian) had been one of the first books published just before the holidays that year. We gave it to my mother for Christmas. That book was the busiest item in our family. The minute somebody put it down, another would pick it up to read. As we drove through the South, I excitedly pointed out to the Benders how this scenery, the Spanish moss, and the plantations were exactly as I had imagined them when I had read the book many years ago.

We stopped for the night in Roanoke, Virginia. In the room that night, I took out every second stitch from my thyroid operation's neck wound using the bathroom mirror.

The next day we got to Houston, the city of the future. It was exciting for me to see the city and the sprawling medical center with its hospital's medical schools and research institutions. We stayed at the Hilton Hotel, which was made quite famous for its distinguished Texan guest list. A major part of the movie *Wild Cutter* was filmed there. This was where the famous Mr. McCarty rode his horse up in the elevator to his suite. To me, what mattered more was its proximity to the medical center.

I had my first interview the next morning at 10 o'clock with Dr. Jim McMurry, who was a Harvard graduate and one of the most outstanding young Americans. He was kind, polite, and very thorough. He concluded by saying, "I think you ought to talk to Professor Jordan, Director of Training Programs. This was great news as it meant I had passed the interview. Otherwise he would have just walked me to the door and wished me a nice trip back to New York.

Dr. Jordan came from the Mayo Clinic and was an author and world expert on diseases of the pancreas. The interview with him was a little tougher. One of his questions was the following, "What would you do if someone was brought into the emergency room with a serious poisonous snakebite, and you were there alone?" I was pretty sure I had flunked the interview with my answers, but he kept on going. When he was finished, he said, "You ought to see Dr. Cooley." This was a good sign, because he was the best living heart surgeon in the world and number two in the department of surgery.

After I had waited for more than an hour, Dr. Cooley asked me quite a few questions about operas, which was great, because it was familiar territory. I used to attend operas in Budapest, and my brother was an opera singer. Furthermore, the top-floor tickets at the opera were cheaper than a movie, so we went to the opera to warm up.

Dr. Cooley asked me if I found it unusual for an American to talk about operas. He also said that he had served in the U.S. Army in Vienna in 1945. At that time, opera was one of the few entertainments in Vienna, and that was why he had become so interested.

About my application, he said, "If you want to come here, we would be happy to have you." I asked him if he had seen my grades, for they were not very good. He said that he had seen them, but if I could pass medical school with my limited command of English, I must be extremely smart. Furthermore, all of my professors thought I was a very hard worker. Then he said that I ought to see Dr. DeBakey, the world-renowned chairman of the department. I couldn't believe it. Less than four years after I had arrived in this great country as a penniless immigrant; I was getting ready to graduate from a very good medical school, and *I had been accepted* by one of the best surgery programs in the world where only the very best dared to apply.

Dr. DeBakey was very busy, and I waited until 9:00 p.m. before he could see me. In the meantime, the Benders had expected me back in the hotel by noon. I could not see any accessible phones in the waiting areas to let them know I was running late, and I was afraid to leave and look for a phone in case I was called. For the same reason, I was afraid to go to the restroom all day.

Finally, I was escorted to Dr. DeBakey's office. I felt like the young novice who had been invited to see the pope. He was reviewing something with extremely focused concentration. Without looking up, he waved at me. After a while, he looked at my dossier and said, "Dr. Makk, we will be glad to grant you a fellowship. I will see to it that you will get enough of a stipend to save some money to take back home with you." Then he got up and walked me to the door. All I could say was the following, "Thank you, sir. It is an honor to meet you." Here I was, a senior medical student applying for an internship. Fellows are doctors who have already finished their surgery training—and in this case, heart surgery training—and then pursue a program such as this to refine their expertise.

At the Baylor surgery program, there were fellows from all over the world. I was not about to tell Dr. DeBakey that I was applying for an internship and that I was not going back home. I wondered how much he had looked at my application, but before I could say anything, I was out of the office. Nearly all the interns and residents were Americans without accents, graduating from the best medical schools, whereas many of the fellows came from other countries with thick accents. This could be the reason he had considered me as a fellow. I took a deep breath and hoped for the best. After I found a restroom, I ran back to the hotel, where the Benders were anxiously waiting for me. They had been waiting so long that they were about to call the police to report me as a missing person.

Next day, I had a chance to meet with some of the leading surgeons in Houston through the kindness of Dr. John Wall, an old friend of the Benders. I decided to list Houston as my first choice and Bellevue as my second for the matching program. I still don't remember if I flew back to Albany or if I returned on cloud nine.

Next came some storm clouds on the sunny skies. Dr. Frawley, the endocrinology professor who had treated my thyroid disease, got a hold of me and chewed me out for having my thyroid operation. I tried to explain to him that for more than a year, my medical treatment had not worked and that I couldn't start my internship with it, but it was to no avail.

One of my best surgical rotations as a senior in medical school was Dr. Stewart Welsh's service. He was an outstanding, nationally known surgeon and quite a character, too. He usually had his medical students chauffeuring him to consultations. Besides being an outstanding surgeon, he was a good eater and drinker. On my first day, we got through an eight-hour operation, and Dr. Welsh threw his car keys at me and said, "Makk, go and warm up my car while I change, and wait for me at the front entrance of the hospital."

I told him that I would be happy to try, but I didn't know how to drive. He blew up and asked, "How did you get on my service? You don't drive, and you don't even speak English."

There wasn't much I could say. I became the first medical student whom he was driving around. But things improved, and one day, he took another medical student and me home for dinner. He was a widower, and his cook always had top dinners like steak and roast beef. He always served good drinks and lots of them. After dinner, he called cabs for us, gave us fares, and even slipped us two dollars extra for the tip. He made it clear that he expected us to tip the cabbie well, because he didn't want any cabdriver thinking he had cheap guests.

After surgery, I had my last internal medicine rotation. One of my patients had double pneumonia and a brain infection with one of the worst germs called staphylococcus. He couldn't cough up his slime, and he needed a tracheotomy. When we were getting ready to start, the thoracic surgery resident gave me the scalpel and said, "You do it." I asked him if he was sure and he said yes. He said he had seen me working in surgery and knew I could do it. I tried not to let my hands shake, and I was able to do it. This was a big deal, because medical students ordinarily didn't get to do tracheotomies. The patient needed his tracheal excretions suctioned out every fifteen to twenty minutes. Now we could do it through the tracheal tube. The lung infection was brought under control, but the brain infection became worse and would not respond to the intravenous medications. My professor decided to drill small holes into the patient's skull and give medications directly to the surface of the brain. Because we needed to administer the medication on time and perform the tracheal suctioning as needed right away, there was no tolerance for delayed medication or suctioning. When I was away from my critically ill patient, he got worse; so

I put an easy chair next to his bed and stayed there for several days, giving the medication and doing the tracheal suctioning to keep his airway free and to remove the accumulated puss in his airway. Finally, he slowly started to recover. It was a great day for me when he walked out of the hospital, recovered. I watched with tears in my eyes as he was leaving. Suddenly, I realized all my setbacks and sacrifices to be a doctor were worth it. That was what medicine was all about.

In the meantime, we received our intern matching results. I got my first choice, the Texas Medical Center. It was like a dream come true, but my joy was short-lived. I was paged at the hospital, and Dr. Cranston Holman from Bellevue Hospital was on the other end of the line. He let me know in no uncertain terms that I hadn't listed Cornell as my first choice as we had agreed. I tried to explain to him what had happened, but he was not interested. Then he said, "If you drop Houston, we would still take you." When I told him I wanted to stay with Houston, he slammed the phone down. I had no regrets about it, because his actions were not according to the rules of the intern-matching programs, which forbid wheeling and dealing.

This wasn't the only challenging problem I had to face. I dated a very pretty and young nurse named Audrey, and when I went to tell her about my great internship, she was happy but asked, "What about us?" I had made it clear to her that she was free to date anyone. I had also made it clear that I had planned to spend the next few years focusing on my surgical training. Then I told her if we still cared for each other after a year or two, we could continue our friendship. I thought it was a nice way to say goodbye, but she didn't take it that way. She said it was the nicest proposal she had ever hoped to hear. I made it as clear as I could that it wasn't a proposal. No, it was goodbye.

As the end of medical school neared, we were all very happy. One day, I received a page from Dr. Stein, who wanted me to come to the morgue. There, he asked me to identify the person on the autopsy table, because the wife was incapable of doing so. It was Dr. Edwards, dead at age 46 from a heart attack. I was shocked, and to this day, his occluded coronary artery is still clear in my memory. My great benefactor, who had given me a decent job, a chance to co-author an important medical publication, a man who had wanted me to get a doctorate and stay in research, was now gone. It was a great loss to science and to me personally.

In the meantime, graduation approached. After all my troubles and sacrifices, I was going to become a doctor but not just any doctor, an

American doctor. Our graduation was at the end of May, and we were to start our internship on July 1st. I couldn't get a job as a technician or research assistant anymore, because I was now a doctor, and I couldn't find a job as a doctor for one month, because I had no clinical experience.

Chapter XIII

Graduation and America the Beautiful

Graduation was absolutely great. Finally, I had made it. I was not only a doctor but an *American* doctor. In addition to my medical degree, I also received the Lamb Award, which was given to the graduate who had taken best care of his patients. The Benders came up from New York, and it was a very emotional moment for all of us. Here I was, three and a half years after I had arrived as a penniless refugee, with no place to sleep or eat, and I became an MD. I had also been accepted for internship in one of the best places in the world.

I was very grateful for the Benders' kindness, as they had taken me into their home with loving and generous hearts. Now that this immigrant had not only become an MD, but also had received an award, the Benders gave me an 18k gold Tiffany wristwatch with "Laszlo Makk, MD" engraved on it. I was in heaven and thanked God and my guardian angels for their help and guidance.

Three of us decided to tour the country, starting in Albany, going to the west coast, and then returning via Denver and splitting up there. Patrick Hagihara, my roommate and mentor, got his internship at the University of Minnesota under the world-renowned Dr. Wagensteen. George Sardina, my other classmate, was going to the West Coast. Our aim was to see the country and to visit as many medical centers as we could to see if we wanted further training anywhere else. Patrick had a small English Vauxhall, and George Sardina had a beat-up American car. We planned to leave right after graduation.

We bought sleeping bags, a cooler, and other accessories. I was afraid to take my medical degree with me, because I didn't want to lose it in a car accident or fire, so I asked Audrey, the nurse I had been dating, if she would keep it for me and mail it to me once I knew where I would be living.

On the trip, our plan was to sleep on the roadside in sleeping bags, and every time we arrived at a medical center, we would clean up and shave in the intern's quarters. We would then visit the facility, get an interview, eat a meal in the cafeteria, and then move on. On the road, our main staple was milk and bread for breakfast, lunch, and dinner.

Our first major stop was Detroit, where we visited the Ford Hospital, an excellent teaching facility. Then we went to see the Ford automobile plant. Both were very impressive. Then we drove over to Canada, which was a thrill. Then it was onto Chicago, where we arrived in the late afternoon. We were driving around, looking for a place to sleep on a very cool evening when we finally found a park by the lake after dark. A group of young people was having a picnic with a bonfire, and they invited us to warm up by the fire. When we explained who we were, they insisted that we eat and drink with them. We happily obliged, and after a while, we found a quiet place in the park to roll out our sleeping bags and go to sleep. We woke up to someone pointing a flashlight in our faces and gently kicking our feet to wake us. It was a policeman who wanted to know what we were doing there. He said that no one was allowed to stay in the park overnight and that we would have to move on. After we visited the Chicago clinics, stopped by the Cook County Hospital, and did a little sightseeing, we headed out West.

Before we had left Albany, we had worked on the logistics, namely how far we should travel each day. It was a tight schedule with no tolerance for delay. We were headed west out of Chicago through beautiful countryside with great sunsets. Somehow, we ended up on a narrow country road with an old man in front of us in a big Buick going about twenty miles an hour. At one point, Patrick felt we could pass him and gunned the little Vauxhall. Just as we were passing him, the old farmer decided to make a left turn and hit us really hard. Our car tumbled for a while. The cooler flew through the windshield, and as our car was rolling, it took out a mailbox. We had glass fragments all over us and a few bumps and scratches but no broken bones. The Vauxhall had scratches and dents, and it now had no windshield. George Sardina's car was okay. The family whose mailbox and fence we had destroyed was most helpful. We offered to fix the fence if they gave us the tools, but they wouldn't hear of it. In the meantime, it was getting dark, and we asked if we could sleep in their yard. They wouldn't hear of that either and insisted that we sleep in their half-finished basement that had a shower. They were fascinated with our story.

The next morning, they cooked us breakfast, which was a feast of pancakes, eggs, sausage, and potatoes. The man of the house also told us where there was a body shop about thirty miles away, where we could try to find a replacement windshield. It was quite a challenge to keep insects out of our eyes as we drove there without a windshield, and I was sure it was quite a sight for the locals. The body shop didn't have a windshield that fit, but they told us there was a large junkyard that might have a windshield for our car. We lucked out and did find a windshield that fit, so we put it on top of the car and secured it with strings, George holding it on one side and me on the other. We were quite a sight again. It was slow-going, because if we had hit a bump, the windshield could have broken. The body shop put the glass in right away, and then we were on our way by midafternoon, trying to make up for lost time.

Our next major stop was Rochester, Minnesota, where we could visit the world-famous Mayo Clinic. There, I had a very good interview with Dr. Clagett, the head of cardiothoracic surgery. I had the feeling then that if I wanted to come there for residency training, they would have allowed me. Then we headed to Minneapolis, which was ninety miles away. We arrived there in the evening. Dr. Walton Lillihei, the world-famous pediatric heart surgeon had just started his rounds and invited us to join his entourage. He treated us like we were on his staff, asking questions and explaining techniques. It was extremely flattering. That evening in the middle of June, there were snow flurries as we crossed from Minneapolis to St. Paul.

Yellowstone, the Badlands, and the Black Hills were our next stops, where we split with George Sardina, because he wanted to do more sightseeing on his way to California. We headed to Salt Lake City and saw the tabernacle and other sights. We decided to cross Death Valley at night, so the trip would be easier on the car and on us because of the daytime heat. We bought the customary water bags for the radiator, filled up with gas, and got halfway before we were nearly out of gas. Finally, we saw a gas station around five in the morning, but it was closed. We managed to find someone asleep in the station. When we woke him up, he told us the gas station was closed and went back to sleep. We decided to fill up anyway and left the money under a stone on the gas pump.

Soon, we were in California and tried to go through the Bremmer Pass. It was snowing, and the trucks were putting chains on their tires; however, Patrick managed to maneuver his little Vauxhall through the snow.

Our next major stop was Sacramento. Patrick had some Japanese relatives there who had prepared a Japanese feast for the new doctor. He asked me if I wanted to go on to San Francisco on a bus because I wouldn't like such delicacies as raw fish. We looked at the map and marked a point on the main road where we would meet two days later. In San Francisco, I found a reasonably priced hotel from which I could go sightseeing for two days. By the evening, the hotel was full of merchant seaman and sailors. Some were not only drunk but fighting, too. My room had a bed, a desk, and a beat-up chair. To secure my door, I pushed the desk against the door and the bed behind it. Eventually, I got some sleep.

San Francisco was fabulous—the Golden Gate Bridge and the hills with their streetcars. After two days, I went to the marked spot on the map. After I had waited for two hours, I started to get nervous. Patrick had most of my belongings in his car, and I started to wonder if he had had a car accident. How would I get to Houston from there? I was very cold, so I kept walking fast to keep myself warm. Finally, Patrick arrived in the little Vauxhall three hours late. He was a welcome sight.

There was one place I didn't get to visit during my first two days in San Francisco, the Top of the Mark Hotel. I had read about it in a magazine—how beautiful the view was from the restaurant on the top floor and that they served the best martinis. I didn't know what martinis were at that time, but it sounded like something good. After we saw how beautiful the restaurant was, we figured we couldn't afford to go in there, but the maitre de asked what he could do for us. We told him about the magazine article we had read about the restaurant, and he invited us in to enjoy the view and a complimentary martini. He would not even accept a tip as we left.

Next, we saw the beautiful communities as we headed south along the Pacific Ocean, Monterey, and Carmel-by-the-Sea. The sea lions we saw are still in my memory today. Then we turned inland to see Yosemite Park. In Hungary, my parents had a volume of *The Book of Science* when I was thirteen or fourteen. There was one picture that somehow had stuck with me, a picture of the giant Sequoia tree in Yosemite Park, showing a car passing through an opening at the bottom of the trunk. I never dreamed that one day I would get to see that tree, but here I was, driving through one of nature's most famous tunnels.

We headed for Los Angeles, but our time was getting short, so after we visited Hollywood and took in a few other sights, we left that same evening. We arrived in Las Vegas around five in the morning, and everything was in full swing just as it would be at five in the evening. We had a very good breakfast for a very reasonable price, and then we played the slot machines for a few minutes. I won a bunch of silver dollars and quit while I was still ahead. Then we left for Denver.

After we had seen Denver, we stopped in a park and divided our belongings. Patrick then dropped me off at the bus station before he headed to Minneapolis to start his internship, and I took a bus to Houston.

The per capita cost of our trip was $182.00 for each person. Our frugality had paid off, and we not only finished this grand tour on time but under budget, with some money left over to start our internships.

Ever since this journey, I have never thought of my new country as America or the United States. I always think of it as "America the Beautiful."

Chapter XIV

Internship Year

The trip to Houston was long. When I had had my interview in Houston a few months earlier, the Benders had introduced me to one of Mrs. Bender's old friends, Dr. John A. Wall, and his family. Dr. Wall was a very distinguished gynecologist. He was Chief of Gynecology at the Anderson Cancer Center and the Director of Obstetrics & Gynecology at Methodist Hospital. Both Dr. and Mrs. Wall were very kind, and he was sure that I would get my internship in Houston. They asked me to let them know if I got the internship.

I did notify them, and they invited me to stay with them until I got situated in Houston. When he found out that I was taking a bus, he insisted that I notify him about the arrival time, because he would pick me up at the station. I was supposed to arrive around 11:00 p.m., But, we encountered stormy weather and didn't get in until 3:00 a.m. At the time, I had no way of letting Dr. Wall know about the delay. I didn't expect him to wait that long, but there he was, with a big smile on his face in the pouring rain, as if it was only three o'clock in the afternoon. We both got soaking wet. He showed me to my room later that morning. At 5:00 a.m., I heard somebody moving around. It was Dr. Wall, and by six o'clock, he was on his way to check in on his patients before he started surgery at 7:30 a.m.

The next day, I got up and went to the hospital with him. While he operated, I walked around the hospital memorizing the location of various wards including the blood bank, the laboratory, and the radiology department. I also got a hold of a hospital phone book and memorized all of the extension numbers. In this way, when I started my internship,

I didn't have to take time to look up phone numbers or ask directions to various locations.

At this time, I also met Jim Moore, a fellow intern and a Cornell graduate. We became buddies right away. He knew a place for rent that was within walking distance from the hospital, which was good, because neither one of us had a car. We shared an efficiency apartment because we figured we would be on duty on alternating nights and would take turns sleeping in the one bed. We went to Sears to buy the cheapest kitchen utensils and other necessities.

In the afternoon, Mrs. Wall very kindly took Jim and me for a swim at the Houston Country Club. There, she introduced us to Mrs. Cooley, who was very nice to us and jokingly remarked that interns were supposed to look exhausted or like somebody who was near death. "Not like you," she said. The next day, June 30th, was orientation introduction and service assignments. My first internship rotation lasted four months with Dr. Denton Cooley's service. It was a great honor and very hard work.

I reported to work at 6:00 a.m. the next day as directed. The other doctors on Dr. Cooley's service were Dr. Hector Howard and Dr. Dick Sinott, both trained heart surgeons who had come for the year, to fine-tune their heart surgery skills. In addition, Dr. Arthur Bell joined them. He was Baylor-trained in general and heart surgery and had just served three years in the U.S. Navy.

Everyone on the service team was very cordial and hard-working. Dr. Bell, my immediate supervisor, demanded perfection with very little tolerance. We finished rounds and completed the takeover from the previous service team by 7:00 a.m. Dr. Bell then assigned the next doctor who would scrub in to assist Dr. Cooley in surgery. I was to be second or third assistant in surgery and would help take care of the patients in Dr. Cooley's service. Patient care involved conducting physical exams, placing laboratory orders, performing X-ray studies, removing chest tubes, taking stitches out, and presenting patient progress reports to Dr. Cooley.

The surgeries finished around six o'clock in the evening, and then we made rounds on our post-operative patients and on those who were scheduled to be operated on the next day. Having the opportunity to work with Dr. Denton Cooley, the high priest of heart surgery, went beyond my wildest dreams. That day, he performed nearly 10 open heart surgeries or other operations. I felt like the novice priest who was not only getting to go to the Vatican but also getting to assist the pope in his daily activities. Then around nine o'clock, Dr. Cooley left, and we finally sat down for coffee. I was given the work list for the rest of the evening and the next day. This included giving exams and doing histories and physical exams on the 12 new patients and scheduling them for diagnostic studies the following day.

At this point, I asked Dr. Bell if we should make a call list. He asked me what I meant. I asked who was going to be on call and spend the night in the hospital on what day. He looked at me and said, "Drs. Howard and Sinott are fellows in heart surgery, and I took calls for four years in general surgery, three years in heart surgery, and three years in the U.S. Navy. Your assignment is for four months, and you will be on call every day and night. But if you ask stupid questions like that, you won't last more than four days." I just couldn't imagine any human being capable of handling that kind of a schedule. After the others went home, I finished writing histories, physicals, and orders for the next day and completed studies on all new patients. It wasn't until 1:00 a.m. that I found my bed in the call room. I had a few calls through the evening and got up at 5:30 a.m. to work another full day and then some. Thank goodness they kept razors and shaving cream on hand in the call room.

Realizing I most likely would not be going home for four months, I asked my roommate Jim Moore to bring me some of my belongings. Once my body was used to sleepless nights, I enjoyed every minute of this busy and outstanding internship rotation. Dr. Cooley seemed to like me, and after a while, even Dr. Bell expressed satisfaction with my work. The service experience itself was tremendous. I got to assist in operations that were done for the first time in history.

One such operation was putting an aortic valve into a patient's aorta that was in Stage IV heart failure. She was 24 years old. She had four children and her husband had left her. Dr. Cooley explained to her the risk of the never-before-performed operation, but she had no choice. This was her only chance to live. Just before we wheeled her into the operating room, she gave me the picture of her four young children and said, "Dr. Makk, you were always very nice to me. I don't have anybody else to ask. If something happens to me, would you look after my children?"

I said, "Yes, of course." Then I realized that if she didn't make it, I would have to take care of four young children on my $75 a month intern salary. I assisted with the operation. Dr. Cooley did a miraculous job, and the patient eventually recovered, even though the operation held great risks because of her advanced heart failure.

One evening during rounds, Dr. Cooley told her that she was ready to go home. She gave Dr. Cooley her insurance forms with the comment that she knew they wouldn't cover his fees but that she would make every effort to pay the rest of the bill. Dr. Cooley signed the insurance papers and told her that whatever payment she got from the insurance company, she should use it to buy something for her children for Christmas. "No bill from me," he said. It made an everlasting impression on me. I thought I was working for a saint who was also a brilliant surgeon.

Dr. Cooley had many challenging patients referred to him, some by other heart surgeons who didn't think they could handle their cases. One such patient was a woman who had been referred to Dr. Cooley by Dr. Paulson, the leading heart surgeon in Dallas. She had a heart tumor as big as her heart. Studies indicated that if it was not removed, she would die soon. Its removal necessitated severing a major coronary artery, which could result in her immediate death. However, Dr. Cooley never hesitated. The operating room was very tense. Everyone was holding their breath. Dr. Cooley cut the artery, and the heart didn't stop! Just as the tumor was removed, Dr. DeBakey walked into the operating room with a visiting Soviet heart surgery professor. They just marveled about the case. Then Dr. Cooley told Dr. DeBakey: "Mike, tell the professor he should have seen what I removed last week. It was twice as big as this." Of course, he never had such a case. It was just to make the Soviets jealous.

After I started my internship, the Benders sent me a newspaper clipping from the Albany paper. There was Audrey's picture and an announcement of her engagement to Dr. Laszlo Makk. The Benders were surprised and expressed some indignation. They said that I should have let them know about it. Naturally, I was surprised, too, as I thought I had said goodbye and good luck. I had never proposed to her, and I had made it very clear that we were not engaged.

I was on call every day and night for the first four months. Audrey kept calling me at the hospital, often on Saturday nights, indicating how much she missed me and wanted to come for a visit. I had no free time, because I was in the hospital day and night, and sometimes her calls interrupted medical procedures I was doing.

One Saturday night, I was working with an 18-year-old boy who had been in a car accident which resulted in his windpipe being torn in two. He was bleeding heavily into his lower windpipes and choking on his own blood. I had a tube down his windpipe and kept suctioning the blood, trying to keep it open, while the operating room was readied. Then I was paged, and I asked one of the nurses to answer my page. She said it was a long-distance call and the caller wouldn't give her name but said it was urgent. I told her to take a message and tell her I would call back when I could. There was no message or call-back number given, but a few minutes later, this was repeated. A while later, I was paged—*stat*. That, I thought, I had to answer, because a timely answer to a stat page in a hospital might save a life. Instead of a true emergency call, it was Audrey on the line again, upset that I hadn't answered her calls. She sounded like she had been drinking, and she kept repeating herself. Finally, I just told her I had to go and to please not call me again. (This time she honored my request, and I never heard from her again.) In the meantime, our young patient survived the transfer to the operating room. The surgery by Dr. Bell was brilliant, and the patient recovered.

We operated on and took care of hundreds of patients. The memory of some patients has faded, but there were many unusual patients whose cases are still with me. As I learned more, I became more and more skilled which led to my increased involvement in patient care. This was before intensive care units existed. Along with about 40 other patients going through a typical recovery period, we kept our really sick patients in the recovery room for days or even weeks. At times, I got so busy that I couldn't leave. The nurses who helped me through the nights set up a gurney with a blanket so that I could get a little rest. They would wake me up when they needed me, wheeling me from one patient to the next as needed. I would try to get a little shut eye in between caring for patients.

Dr. Cooley liked my work and gave me more and more responsibilities. He also noted that I was getting a little pale. One Saturday morning, he took me to his ranch to get some fresh air. I had never ridden a horse before, but I ended up riding around with Sir John Barrett who was his good friend from Australia. He and Dr. Cooley had trained together in London. (The condition "Barrett's esophagus" is named after Sir John Barrett.)

Near the end of my rotation on Dr. Cooley's service, we had just finished an open heart operation and the heart pump machine was connected to the patient through the femoral artery. It needed to be sutured, which was the job of a fellow or third-year resident. Dr. Allen Lansing was a fellow and was supposed to sew up the artery. Instead, he motioned to me to do it. Sewing arteries was very taxing, because unless it was done perfectly, the pulsating pressure of blood would pull the stitches apart and cause catastrophic bleeding. My heart was in my throat, but it went well. (Interestingly, Dr. Lansing would end up establishing his career in Louisville, Kentucky as did I. He became Louisville's leading heart surgeon and was known throughout the world for his work in the field.)

The four-month rotation with Dr. Cooley finally came to an end. (I got to sleep in my apartment for the first time in ages.) I really loved every minute of the experience. I was a different doctor now and battle tested.

My internship consisted of eight months of surgery and four months of electives. Now it was time to decide on my elective rotations. I decided to take two months of nuclear medicine, a month of radiology, and one in gynecology. Nearly all the surgeons were outstanding teachers.

Dr. John Overstreet was an outstanding surgeon as well, very arduous and slow in the operating room. He kept unusual hours, starting operations late, going to the office late, and starting evening rounds around seven or eight o'clock, whereas most other surgeons finished by six, and I would have time to stop by the library on my way home. But when I covered Dr. Overstreet's patients, I wouldn't be out of the hospital until 10 o'clock on most night. He was a great teacher, and I kept learning from him, so I really didn't mind what time I finished the day.

One of our challenging patients was a Mrs. Smith, who had had all four of her parathyroid glands accidentally removed during a previous thyroid operation. She suffered from many complications as a result, including severe headaches as well as calcium and muscle problems. She needed daily injections of parathyroid hormone. Her headaches used to come on in the evening, and I would get a call from her nurse, asking if the patient could have two aspirins. I made it my custom never to give medication orders over the phone, always checking the patient first even if it were just an aspirin or two. Every night I went to check on Mrs. Smith, Mr. Smith would be there reading to her, devotedly trying to keep her focus away from the pain. Each time, they both would apologize profusely for bothering me. She got better slowly, but I worried about how she would get her daily shots. I decided to teach Mr. Smith how to give her the shots. This was during the "old days" when we had to boil the syringes and needles. I also showed him how to file needles so they wouldn't get dull and hurt more. On the morning of her dismissal, there was an exhibit of disposable needles and syringes (the first time I had ever seen them). I told the detail man to give me a bunch of them. I ran up to Mrs. Smith's room and showed them both how to use the syringes, and then I left. I was never one to hang around patients at discharge time to avail myself to their expressions of gratitude.

About one month later, I got a very nice invitation from the Smiths for five o'clock tea. I didn't know where the address was, but one of the doctors from Houston said that it was one of the big mansions near Rice University which was nearby. Everyone said I should wear a necktie and jacket and should not try to walk there. Luckily, Bob Andrews, a resident physician, loaned me his car.

There was a guard booth before the entrance, and the guard directed me to the place. As I pulled up to the house, an Irish butler wearing a tuxedo opened the door. He was holding a silver tray out for my calling card, but I didn't have one. He said, "Dr. Makk, Mr. and Mrs. Smith are expecting you."

The other guest for tea was Sir John Barbirolli, the famous conductor of the London Symphony Orchestra who was guest conducting for the Houston Symphony. I had been on call the night before and slept very little, but I managed to participate in the conversation.

I was invited again, this time to their New Year's Eve party. Everyone thought I should wear a tuxedo, but I didn't own one, so another resident physician, Tim Lowery, loaned me his.

On New Year's Eve, I ran into Dr. Hans Altinger who had just finished his gynecology training that day and was ready to start private practice the next day. We agreed we should celebrate and that martinis would be the best way.

We went to my place first, so we decided to have a drink there. I hung up Tim Lowry's tuxedo on the shower curtain rail. We only had water glasses, so we made our martinis in them. I don't know what happened next, but I woke up around midnight, fully clothed in the bathtub, staring at the tuxedo. Then I realized I was supposed to be at the Smiths' New Year's Eve party around 9:00 p.m. When I woke up a bit more, I called them to apologize for not showing up, and Mrs. Smith said, "We know how busy you doctors are. Come over now. We have your dinner in the oven. The party will be going for quite a while."

So I cleaned up and got a cab.

After I arrived, I was introduced to many of the old-money elites of Houston. Everyone was very nice and fascinated with my life story, and they all had tremendous respect for my being trained in the surgery program. I left around 3 a.m. with the party still going on. I caught a little sleep, and I was back on call at the hospital shortly after 7:00 a.m. the next day, really sleepy.

I was hoping for a light day, but it turned out to be quite busy. I remember one patient in particular, a man who had inhaled a peanut into his windpipe at a cocktail party and was now choking on it. Because he had had a few drinks, he couldn't be put to sleep. Under local anesthesia, Dr. Overstreet put a steel tube into his windpipe and tried to extract the peanut with long forceps. It was a very delicate job, for he had to grab the peanut without breaking it. If it were to fragment, pieces would scatter all over the lungs, and their high salt content would cause a terrible pneumonia. As he kept grabbing it with the forceps, it kept slipping away and got further down into the bronchial tube. My job was to hold the head steady and securely, because any movement of the head could produce great complications. I never realized how heavy the head could be. Finally, after two hours, the peanut was out.

My next rotation was in nuclear medicine, and it would affect the rest of my life. It was an exciting new part of medicine. Dr. Philip Johnson, the director, had developed a technique where one could measure the blood flow to the coronary arteries in the heart. One of my professors had an U.S. Army grant to study the coronary blood flow in patients who were in shock. We eventually put a project together during which I was to measure coronary blood flow in trauma cases while the patients were still in shock but after their treatment had been initiated. Every Saturday night when I was not on call, we packed up the heavy, bulky instruments from Methodist Hospital and transported them to the big trauma center at Jefferson Davis Hospital. After we set them up, we would wait for the right patient. Around 3:00 a.m., we would take the equipment back to Methodist Hospital. We continued this study and eventually observed previously unknown information on coronary blood flow in shock patients, which was eventually published in a prestigious medical journal.

My nuclear medicine rotation yielded another benefit that lasted through my lifetime. And that was meeting a beautiful young lady who worked in one of Dr. DeBakey's research laboratories on the same floor. Carolyn Cooke was a beautiful, kind, and smart girl who was always impeccably dressed in her white lab uniform. Finally, my chance came when we got stuck in the elevator together. Both of her hands were full of lab glassware in racks. I tried to assure her that everything was going to be okay. She was polite but in no mood to strike up a conversation with a stranger.

On a later occasion, I was introduced to her, but I was in no position to ask her for a date. I didn't have a car, and Houston was such a car-oriented city that many streets didn't even have sidewalks. Then I bought my first car. It was a used 1957 MG sports car. I didn't really know how to drive, but I practiced driving when I could. During that time, Dr. Cooley asked me if I would like to go to the rodeo. He gave me the tickets to his box. I now had a car *and* tickets to the rodeo on Saturday. I went up to the lab where Miss Cooke worked and asked her if she would come to the rodeo with me. She said yes!

I picked her up in my MG and quickly realized the difference between practicing driving on abandoned streets and driving in traffic. There was a hairpin curve by the museum on the way to downtown Houston. I couldn't negotiate it, and my car jumped the esplanade and headed into the oncoming traffic. Everybody was blowing their horns at me. Luckily, I was able to jump the car back to the esplanade and onto the correct side of the street. I was shaken up, but told Carolyn *these sports cars do this every once and a while*. She said, "I don't know about that. I used to date a fellow who had a Jaguar, and his car never did that." I wanted to know who this "SOB" was who owned a car was better than mine.

Overall, our first date went fine. I'm not even sure if I saw the rodeo. I just watched her as she sat next to me. After the rodeo, she asked me if I would like to go to the opera with her that night because she was going to usher there. Of course I said I would.

When I dropped her off, I realized I didn't have any money left, and none of my roommates were home to lend me some. The apartment next to ours was rented by Dr. Arthur Page, a surgical fellow and later a distinguished Canadian heart surgeon. Luckily, he was home. I asked him if he could loan me five dollars, because I was supposed to be taking this wonderful girl to the opera. He said he would loan me 10 dollars if I would have a drink with him and tell him all about her. I kept talking about her and got my 10 dollars. The opera that night was great, but I spent most of my time admiring her.

Everything started to go my way. All my surgery chiefs liked my work now. I was working on two research projects in addition to my regular duties. One project was on the ruptured sinus of Valsalva for Dr. Cooley. The other was on coronary blood flow for Dr. Johnson. In addition, I now had Carolyn in my life.

On my gynecology rotation, Dr. Wall had a patient whose hysterectomy wound pulled apart. Not only did it not heal, but it kept getting worse with a lot of tissue in the abdominal wall becoming necrotic, regardless of the treatment the outstanding consultants recommended. I changed the dressing several times a day and removed the newly formed dead tissues. One thing I noticed was that there were always one or two well-dressed men hanging around her room, regardless of the time of day. Finally, she got better. When she was ready to be discharged, Dr. Wall told her husband that if it hadn't been for Dr. Makk's extraordinary effort, she probably would not have made it. Her husband wanted to know about my accent. On leaving, he gave me his card—Oscar Hawkins, Director of FBI, Houston.

Later, Mr. Hawkins invited Carolyn and me for dinner. Oscar told me the men by Mrs. Hawkins' room were FBI agents. He posted them there because all he knew was that she was getting worse, nearly terminal, and that there was this young doctor with a strange Eastern European accent coming and going at all hours of the day. Of course, that young doctor was me.

We became good friends. One day, I got a call from him. He wanted to know if I would like to go deep sea fishing for three days. I told him I was working and couldn't get that much time off. He said this would not cut into my vacation time and wanted my chief's name. Next thing I knew, Dr. L.L.D. Tuttle, the Chief of Surgery, called me in and said the head of the FBI had called and told him they needed me for three days. "Good luck and go," the Chief said. I was sure Dr. Tuttle thought I was on a secret mission instead of deep sea fishing.

The boat was an old minesweeper converted into a luxury yacht, which slept 20 in air-conditioned quarters. There was a bar in the midsection that occupied the full width of the boat. I decided I would taste all of the drinks in the bar, but there were so many that I couldn't even try half of them within the three days.

The boat belonged to the Brown & Root construction company, and this trip was for law enforcement officers, judges, and police chiefs, A band was playing as we were taking off. The fishing was fantastic, and the largest fish we caught was a 267-pound grouper. The food and company were excellent as well.

When I arrived back at work, the Chief of Ophthalmology and a professor of plastic surgery operating on a child needed an assistant. I was free, so I volunteered. The child had a non-cancerous nerve tumor, which grew in irregular bundles around his eyes. Called plexiform neuroma, it looked like a bucket full of snakes. The operation went well, and when I had time, I went to the medical library to research this rare tumor. Then I presented my findings to the surgeon who asked me to write the case up.

It was presented to an eye conference and was published in a prestigious medical publication afterward.

During June, the last months of my internship, I worked on one of Dr. Tuttle's patients whom I thought had appendicitis. Interns were only supposed to assist in surgery, but before the operation commenced, Dr. Tuttle asked me to come over to his side at which point he gave me the scalpel and told me to perform the operation. There I was, doing an operation as an intern, on an open abdominal cavity.

As my internship was coming to an end, I got a page from Dr. Frank Ashton, a British heart surgeon calling from the library and congratulating me on the nice article that Dr. Edwards and I had written when I was still a medical student. It turned out that the article had just been published in the *American Journal of Pathology*.

This was the most glorious year of my life, but there was more at the closing of my internship. I received a prestigious prize, the Outstanding Intern of the Year Award. I couldn't believe it when I heard my name called as the recipient, and I really had to try hard to keep myself from crying. It was a beautiful way to finish my intern year.

Chapter XV

The Making of a Surgeon

In July of 1961, I started my surgical residency with Dr. Michael DeBakey's surgical program at the Baylor College of Medicine located in Houston,Texas. During our orientation, Dr. Jordan, Director of Surgical Residency Programs, said there were 27 of us first-year residents, but only four of us would finish. I thought, *Welcome to hell*, but this was what it took to achieve excellence in surgery.

The residency rotation in Dr. DeBakey's service was the crown jewel of surgical training programs. Patients were sent to him from all over the world. Other heart surgeons referred many cases to him because they were too complex for them. He also had fellows who were already heart surgeons come to his program from all over the world to fine tune their skills. Dr. Howell, one of DeBakey's associates, gave me some advice in good faith. He said that Dr. DeBakey liked to operate on patients referred to him, and he did not like contraindications to surgery. For example, if a kidney test called BUN was 62 (normally 10 to 20), you would have to put a one in front of it so that it would read 16.2. Then you would have to go to the lab at night and do the same on the lab copy of the test. I thanked him and tried not to show my dismay, but decided I would never do that.

One of our surgery professors was very demanding, but he appreciated my hard work. Before I was assigned to Dr. DeBakey, he took me home for lunch one Saturday and told me I was doing very well, but I would not finish my rotation with Dr. DeBakey, because he never let a foreigner complete his service.

I was not a "foreigner" for long. I became a U.S. citizen in 1962. It was a most happy occasion. Many of my professors came to the ceremony, and the Saturday morning surgical conference was replaced with a surprise party for me. Everything was going great. I decided to propose to Carolyn Cooke. She accepted.

During the week of our wedding, Dr. Tuttle, Chief of Surgery at Methodist Hospital, took me to lunch. I could hardly force the food and drinks down. On our wedding day, we went to early mass, and then I took a nap, because I was very tired. When I was getting ready for the wedding, I noticed that the whites of my eyes were turning a bit yellow.

Carolyn was very sweet, kind, and beautiful. When she walked down the aisle, she was the most beautiful bride in the world to me.

At this time, one of my Hungarian refugee friends was the doorman in a hotel/condominium complex. His wedding present to us was the use of a suite where we could spend our first night as man and wife. I woke up in the middle of the night and went to the bathroom. Now the whites of my eyes seemed to be even more yellow, but I rationalized that it was because of the fluorescent light. What scared me more was that there were toothbrushes in the bathroom, pajamas hanging from the doors, and suits and dresses hanging in the closet. It was obvious to me that people lived there. My first thought was to get out of there before the owners returned home, but I didn't want to start our first day of married life dragging Carolyn out of bed in the middle of the night and wandering around looking for a place to sleep. Therefore, I got up really early to get us out of there in case the owners showed up, and we took off. At a later date, my doorman friend came to visit me, and I told him how concerned I had been about staying in that room that night. He said, "I should have mentioned to you that the suite belonged to an oil company and they only use it for visiting executives. They always call me to let me know when someone was coming to stay."

We did go to Galveston, Texas, for our honeymoon. By the evening of our first day there, I started to turn really yellow, and it became obvious that I had hepatitis. The next morning, instead of heading for the beach, we drove straight back to the hospital where I was admitted immediately. My hepatitis turned out to be quite severe. It was at that time that I remembered I had given open heart resuscitation to a 16-year-old boy who had a cardiac arrest and a broken rib. His broken rib pierced my forearm, and he eventually died from hepatitis. I asked the Lord to let me survive it.

My progress was very slow for a while. I was drifting in and out of consciousness. Carolyn worked in the same hospital. One day when she came to visit me, she walked by Dr. Tuttle and two other surgeons and overheard Tuttle saying, "Too bad about Makk. He was such a nice person and a good surgeon." She thought the comment meant that I had passed away. She ran into my room, really concerned, and kept asking me if I was okay. Thank God I was! I was in the hospital for a month and then at home for nearly two months.

The residents on Debakey's service were not expected to leave the hospital during their rotation. The residency rotation took place over three months, not four. I was to remain on call 24 /7. His patient volume was staggering. He usually had close to 100 patients in the hospital. One of the two residents would take care of the floor patients, which involved examining all new admissions, ordering, evaluating, and presenting all diagnostic studies, taking care of all postoperative patients (e.g., wound care, diet, and postsurgical studies), and providing instructions for care and follow up. The resident in intensive care was responsible for immediate postsurgical care which involved medications, diagnostic studies, transfusions, and breathing machines. One of the residents before me lost 36 pounds on this service.

When I was on the floor, I got wonderful cooperation from all the caregivers. I was their boy, the outstanding intern and proud new U.S. citizen. Dr. DeBakey could be the most charming and considerate person with patients and the toughest, cruelest critic with his underlings. He was famous for firing residents on a whim. I survived the first six weeks on the floor fairly well.

On Sunday afternoons, Carolyn would bring me cookies, clean clothes, and pick up my dirty laundry. Sometimes, I was so busy that I didn't have time to even say hello to her.

One of our patients was a wealthy car dealer upon whom Dr. DeBakey had operated and who came back every year for a follow-up evaluation. He always brought a bunch of car brochures with him and would ask Dr. DeBakey to choose which one he wanted and he would send it over to him. This particular time, all of his studies had been completed, and he was ready to be seen by Dr. DeBakey. As we were seeing patients, Dr. DeBakey suddenly announced that he was leaving. I made the mistake of telling him that the car dealer from Beaumont was waiting and expected to see him.

Dr. DeBakey raised his voice and told me not to tell him what he had to do. I tried again to get him to see this VIP patient, but to no avail. Dr. DeBakey had left abruptly. One of his senior associates had to call him to come back and see this very important patient who had given him a Chrysler Imperial. Upon Dr. DeBakey's return, he had very unkind words for me about why I hadn't told him about this patient. Obviously, he just wasn't listening to me.

The next patient that caused tension between me and Dr. DeBakey was a member of the Texas Rangers (a law enforcement division). He had been admitted and waiting for days to be seen by the world famous DeBakey. One morning, this patient was listening to Dr. DeBakey's congressional testimony being given in Washington, D.C. about stroke and heart disease. As he walked by the nursing station, he saw that Dr. DeBakey had eight operations scheduled for that same day. He asked me how that could be possible. I told him I didn't have an answer. Dr. DeBakey finally saw the Texas Ranger a couple of days later when the patient told him "he must be like God". How could he be operating on eight patients here in Houston while testifying in Washington at the same time? As we left the room, Dr. DeBakey turned to me and firmly ordered me: "Don't ever let a patient ask stupid questions like that."

"Yes sir" was my only answer.

Another interesting patient was a rancher from Montana. Dr. DeBakey had told his referring doctor that once diagnostic studies were made, he would operate, and the patient would be on his way home in two weeks or less. The man had no insurance and had already waited two weeks without seeing Dr. DeBakey. Finally, we reviewed his studies. He had an aneurism of his entire aorta. Aortic aneurism was the pathologic dilation of the largest blood vessel. It usually involved only part of the aorta. Its repair was a great surgical challenge, and Dr. DeBakey was very enthused about doing it. However, DeBakey had planned a trip to Belgium and was leaving and couldn't operate on him until after he returned. Because the patient had already lost 30 pounds and had trouble swallowing, I suggested to Dr. DeBakey that he let us do an X-ray study to make sure he didn't have esophageal cancer. Dr. DeBakey blew up saying: "How could you make such a stupid suggestion? A high school student would have better judgment?"

After Dr. DeBakey returned from Belgium, he planned to operate on the rancher. My co-rotating resident, Dr. Quinn had severe bleeding into his windpipe the night before the scheduled operation. It turned out that Quinn had advanced tuberculosis. Until a replacement was found, I was to take care of all patients on the floor and in intensive care. Dr. DeBakey expressed his concern about whether or not Dr. Quinn had infected some of his patients. He didn't appear to be concerned about Dr. Quinn personally or his family.

The rancher was first on the surgery schedule the next morning, but shortly after the surgery started, everything started going downhill fast. The anesthesiologist advised Dr. DeBakey that the patient couldn't tolerate any further surgery. He stopped the operation and sewed him back up before he would have died on the operating table.

I took care of him in the recovery room. He had signs of cardiac tamponade; which would have meant bleeding into the sac around the heart and a compression of the heart, blocking its ability to pump blood. An EKG confirmed my finding. The treatment of choice was inserting a large needle into the pericardial cavity and removing the blood. While we were setting up the instruments, Dr. Alexander, a professor of cardiology, walked by. I asked him to confirm my findings, and he did. Shortly thereafter, Dr. DeBakey appeared. He was very agitated and asked me what I was doing. I showed him the EKG tracing, and he said that I was stupid and that his patients did not have those types of postoperative complications. He picked up the instrument tray and threw it on the floor, shattering the glass parts all over. He picked up a stack of medical charts to make rounds and demanded that I tell him what each patient's hourly urinary output was which was almost impossible to do. As he was looking at a chart, I could read it and told him the information he wanted, so I passed his test. He then slammed the charts on the floor and left for the operating rooms. The cardiac tamponade patient died shortly after Dr. DeBakey left the room.

That day, he had 22 patients scheduled for surgery, all major operations, and many were open heart cases. He had three or four operating rooms going at the same time, with most surgery performed by his associates. Patients were removed from the operating rooms quickly to make room for the next cases. This was an added burden for the recovery room caregivers. One of the patients, who had been given a Dacron artificial artery to bypass the occluded right femoral artery, was brought to the recovery room with a cold right leg that was turning bluish. This meant that the bypass had clotted and had to be redone as soon as possible before he developed gangrene.

I sent word in to Dr. Howell, who had performed the surgery; however, I did not get a response for quite a while, and the patient was getting worse. I sent another message, and there was still no answer. After a third message, Dr. Howell appeared, and after he examined the patient, he agreed with me. He told me that Dr. DeBakey was leaving town and had to finish the surgery schedule, so this would have to wait until tomorrow. But I pointed out that by that time, the leg may have to be amputated, whereas if the patient was operated on right away, the leg could possibly be saved. He raised his voice and said, "You heard what I said," and then he rushed back to the operating room. A while later, Dr. Garret, the senior associate, came out from surgery and informed me that Dr. DeBakey wanted me off his service in the morning. I asked him what I had done wrong, and he said nothing but, "You know how things go when he gets frustrated. We want you back later on to finish your rotation." I told him if I was going to be off the service in the morning, I would leave immediately. He said that I should stay, because Dr. DeBakey often changed his mind by the morning. I was deeply hurt, because I had done nothing wrong, but I was off the service.

I called Carolyn to come and get me. Dr. Garrett said again that I had better stay until the morning because there wasn't anyone to take over my work. I told him that that was Dr. DeBakey's problem, not mine. He was a little shocked. When Carolyn arrived, she brought our infant son. He didn't even recognize me, When I went to pick him up, he pushed me away and cried for his mama.

I didn't sleep at all that night. The next morning, Dr. Jordan, Director of Surgical Residency, called and wanted to know what service I would like to be assigned to next. He indicated that I would be able to finish Dr. DeBakey's service at a later date. I asked him why I had been removed from the service. What had I done wrong? He said nothing that he knew of. In fact, his information indicated that I had been doing a good job. I suppose Dr. DeBakey was frustrated and didn't have a dog to kick, so instead he kicked a resident on his service.

That afternoon, I felt vindicated when I learned that the Montana rancher patient I had worked with not only had a cardiac tamponade, but also had a large esophageal cancer tumor which had spread to the liver (nearly replacing it). I had been right and was proud of myself.

For my next service, I chose "black surgery", which ended up being my favorite. (At that time, there were separate wards for blacks and whites in Jefferson Davis Hospital.) My first patient was a nice 96-year-old black lady who was in heart and kidney failure, and had ascending wet gangrene of the left leg. She needed an amputation, but was too sick and couldn't tolerate the anesthesia. Because the amputation was an absolute necessity, it was decided to go ahead with the procedure without it (while she was awake), and I was to do it. We kept her leg in crushed ice for two days until it was almost frozen and no longer had any feeling. I then cut the leg off above the knee at the demarcation line between living and gangrenous tissue. It was still very painful, but she survived it. Her recovery was threatened by very low blood potassium levels which could have led to cardiac arrest. She also had a rectal tumor which excreted her potassium. The chief resident was called on to remove it. Despite her horrible odds of surviving all of this, she became well enough to go home. It was a great feeling to see my patient leaving the hospital alive.

We residents had a great deal of responsibility caring for many very sick patients. In addition to surgery, we learned about diagnosing illnesses, as well as preoperative and postoperative care. We were expected to know everything there was to know about our patients' illnesses and often stayed in the hospital library past midnight.

I was in one of the most competitive and demanding surgical training programs in the country. I was in seventh heaven.

One of my most challenging patients was an elderly lady who had a disease called necrotizing fasciitis. Starting at her ankle and spreading all the way to her lower back, connective tissues called fascia died progressively under her skin, We had to change her dressing, open the skin daily (sometimes twice a day), remove the dead tissue, and bandage her up again. We buried a tube in the dressing so her wound could be irrigated continuously. The disease kept spreading, but I never gave up. Finally, after two and a half months, she improved to the point that she could go home. Again, I was satisfied, because the odds had been so stacked against her. Consequently, my supervisors noticed my efforts and rewarded me with more responsibility.

My next rotation indicated that I and another resident would be in charge of the emergency room for three months. Each of us were on duty for 24 hours and then off for 24 hours. The one going off duty had to go to the surgery clinic to give follow-up care to the emergency room patients. This was one of the busiest emergency room/trauma centers in the country. In three months, we oversaw more than fourteen thousand patients, but we had excellent nurses and other personnel to help us.

Another rotation was anesthesia for three months. I had to quickly learn how to put patients to sleep or give them spinals. One of our most challenging patients was a 27-year-old woman who was in labor. She also had advanced breast cancer, which had spread to her bones and spine, so she couldn't sit up for the spinal. If we had tried to sit her up, her vertebrae would have been in danger of collapsing. It was arranged that I would give her the spinal with the chief helping. We put two beds parallel placed about 12 inches from each other and laid her across the beds so that the spinal site was exposed between them. I crawled under the beds and gave her the spinal lying on the floor. Within an hour, she had a new baby.

One of my most challenging days in general surgery was the day President Kennedy was shot. It was November 22, 1963, and I was on call. We were all in shock with the news of the assassination, but the emergency room was flooded, and there was much work to be done. Five people with perforated ulcers of their stomachs came to the emergency room that evening and into the night. The treatment at that time was removing part of the stomach and hooking it up to the small bowel. This took about three hours in general. I was going nonstop, and by 7:15 the next morning, I had finished the fifth case. My chief resident jokingly asked why I had filled up the recovery room last night and congratulated me for the new record.

Plastic surgery was another three-month rotation. One of the plastic surgery residents was married to a wealthy Houston socialite and didn't like to take night calls. He told me that if I took his night calls, I could operate on all the patients I admitted or treated in the emergency room. Needless to say, I performed a lot of plastic surgery and learned much.

This particular resident was also a music enthusiast. He moved his elaborate hi-fi into the operating room, and while he was fooling around with it, I went on operating. He became quite famous later, because he murdered his wife before his father-in-law had him killed. There was even a book, a movie, and a TV show about their story of murder and deceit called *Blood and Money*.

Another interesting patient was a lady who presented with abdominal pain. She was brought in from a psychiatric facility to the emergency room. Her abdomen was as hard as a rock, and it looked like gallbladder inflammation; however, the tests didn't support that diagnosis. She was in heart failure, which would have been a great risk for anesthesia. As a last resort, I hypnotized her, and the abdomen softened. I called my chief resident for a consult. He very firmly let me know that in surgery, we would never do that. He called the attending surgeon covering the service, Professor Jordon, the director of surgical residency programs. By this time, it was about six o'clock in the morning. Dr. Jordan made it clear that this was intolerable, and he was coming in himself to evaluate the patient. By the time he arrived, all of my colleagues had disappeared. They probably didn't want their faces associated with the big chew out.

I presented the patient to Dr. Jordan at 7:00 a.m. He examined the patient, and in front of everyone, including the nurses and relatives, he told me that this patient had an inflamed gallbladder, which had probably ruptured and that she was in mortal danger because of my unacceptable care and bad judgment. He ordered the patient to be immediately transferred to the operating room, and decided that he would perform the operation himself. I offered to assist in the operation, but he said absolutely not. I went by the operating room later, and a nurse came out and whispered to me that the patient had no abdominal disease. "You were right," she said. My skin was saved. Afterward, Dr. Jordan said nothing to me about the case, but he was very nice.

Dr. Jordan actually showed his appreciation for me a bit later in an important professional venue. The Dr. Alton Ochsner Surgical Society was meeting in Houston. This was a prestigious professional organization and an important meeting. Dr. Jordan was a member and included my research paper on the ruptured sinus of Valsalva on the agenda, He placed my presentation between the presentations of Dr. Cooley and Dr. DeBakey. I imagine that by putting me in such a distinguished place in the program, it was his way of saying I had been right about my near-career-ending patient with the hard abdomen .

However, my carefully prepared presentation never happened as I planned. While Dr. Cooley was giving his presentation, Dr. DeBakey came in and sat down next to me. By this time, Dr. DeBakey and Dr. Cooley were not the best of friends. Dr. Cooley kept on talking over his allotted time, and Dr. DeBakey kept getting madder and madder about it, telling me how unconscionable it was that Dr. Cooley was eating up his time. I suggested that he take my place, because I could give my presentation later. He said no, but then he asked how long my presentation was. I told him I could make it very short. I got up to the podium and showed my first slide, which defined the disease, and the last slide with the operative results. My paper presented the greatest number of patients ever reported on suffering from ruptured sinus of Valsalva. I finished the presentation in less than two minutes because I was afraid the already-angry Dr. DeBakey might transfer his ire onto me. As I was leaving the podium, our eyes met, and he signaled his approval by nodding his head.

In this surgical program, we had the most interesting cases, as well as the most demanding and brilliant surgeons as mentors. Thankfully, operating came easy to me. At one point, I even tried to teach myself to be ambidextrous. My instructors noticed this and gave me more and more opportunities to operate, even on their private patients. One evening, Dr. Cooley called me and said a patient of his was coming to the emergency room with acute abdominal pain. He said I should examine him, and if I thought he needed an operation, I should take the patient to surgery and do it. He said he would catch up with me later.

The patient turned out to be one of the leading citizens of Houston. I thought he had an inflamed gallbladder, so I took him into the operating room and removed his gallbladder, which was indeed inflamed. As I was finishing up the operation and putting the dressing on the wound, Dr. Cooley showed up. He asked me to go see the family with him. They all thanked Dr. Cooley for the operation, but he told them, "My associate, Dr. Makk, did the surgery. Thank him. I know he did an excellent job." I was in heaven. One of the best surgeons in the world not only trusted me, but also told the family that I had performed the operation. That was real honesty, because residents were, at best, supporting actors in the background.

By that time, Dr. Cooley and Dr. DeBakey had drifted even farther apart, and so had the residents. I was a Cooley man then, and I am a Cooley man now.

In the surgery program, every resident had to spend three months in pathology to learn the "science of disease." When my doctor allowed me to return to work, my liver function tests were still quite abnormal, so I was assigned to the pathology rotation, an easy service compared to surgery, which gave me more time to recover. The assignment was at the 1600-bed Veterans Administration (VA) Hospital where Dr. Bela Halpert, a nationally known pathologist of Hungarian origin, was the chief of the pathology department. Dr. Halpert was a great teacher and a perfectionist.

At orientation, I learned that we would have an "organ recital" every Tuesday morning. I thought that meant we would have a musical recital with an organ. I told Carolyn, "What an elegant program this is." I asked one of my fellow residents what to wear and where the organ recital would be held. He said to wear a scrub suit, and that it was in the morgue where we would review specimens from the previous week's autopsies with senior pathologists under Dr. Halpert's direction. A photographer took pictures of the diseased organs. To avoid glare in the pictures, the specimens were put under water. I learned that technique well and used it the rest of my life.

With my surgical background, autopsies came easy to me. I could do them quickly and well. We once had a very interesting patient with pancreatic cancer. Because he saw my interest in the case, Dr. Halpert recommended I study all pancreatic cancers. When I asked where the registry was, he said I couldn't use the registry. "You need to go through the medical records of each patient who had autopsies and find them." Out of 2000 autopsies, 120 had had pancreatic cancer. At first, I was annoyed to do so much extra work but as I progressed, I realized what a tremendous learning opportunity it was. I reviewed over 2000 medical records from the first symptoms to the meticulously documented autopsy results including photographs—documenting the diseases that caused these patients' deaths. I carried loads of records home for the weekend. Every Monday morning, I had to report the progress of my research to Dr. Halpert. He was super critical of everything, demanded perfection, and I started to get frustrated by his constant criticism. I took a short paragraph from one of his articles he had written on lung cancer 20 years before, adapted it to pancreatic cancer, and included it in my report. Dr. Halpert tore it apart, saying that I was getting worse instead of better. However, I no longer got frustrated. After I finished writing an article on the information from my report, Dr. Halpert admitted that he had always been quite pleased with my work, and just wanted it to be perfect. The article was later published in a prestigious surgical journal.

This work was done in addition to our routine duties which consisted of examining surgical specimens, preparing for conference presentations, and performing autopsies. Juggling all of this could get quite busy. One Fourth of July weekend, I performed 13 autopsies. When the rotation was completed, I had learned a lot, and I came to like pathology. My personal liver tests improved, and I was ready to rotate back to surgery.

My next rotation was with pediatric surgery at the Texas Children's Hospital. Besides being involved with routine operations, we had fantastic cases and brilliant surgeons to train us, and I was allowed to perform major surgeries. One of the surgeons, Dr. Brooks, had a young patient who had a large noncancerous tumor in his chest that was compressing nearly half of his lung and pressing on his heart. While getting ready to operate, Dr. Brooks motioned to me to come to the surgeon's place, gave me the scalpel and said, "You do it." It was a very challenging case and a tedious operation because we had to be very careful not to hurt any of the vital structures in the area. I removed the tumor, and the patient recovered without any complications. It was with indescribable pride and pleasure that I saw him leave the hospital.

Another challenging case was a young girl who had accidentally swallowed lye, which had destroyed her esophagus. She was half the size and weight of a normal child her age because she could not eat solid food. We needed to make a replacement esophagus from the central portion of her large bowel. This was one of the biggest and most challenging operations in surgery. It involved selecting the central portion of the large bowel, cutting it away from the rest with the blood supply intact, making a tunnel between the back of the breast bone in front of the heart and other structures, pulling the cut off segment of the bowel through the tunnel, and connecting it to the uppermost part of the gullet and the lower part to her stomach. I had to make sure not to compromise the blood supply of this segment of large bowel by stretching it too much. If the blood supply was compromised, the whole operation would be a failure, and the patient would miss her only chance to eat again and would end up worse off than she was before surgery.

Dr. Jim Harberg, a tremendous pediatric surgeon, assisted me. As expected, the operation took nearly six hours. Afterward, the patient was recovering nicely, and she could hardly wait to eat her first bite of solid food in three years. She was begging us to let her have a hamburger. We were to present her case to the surgical conference. The day before the conference, I let her have a hamburger. She was the happiest little girl in the world. I had tears in my eyes as I watched her eat.

The following day, when I presented the case to the pediatric surgery conference, I finished off by saying that she had recovered to the point that yesterday she was able to eat her first hamburger. I thought this was my biggest and best surgery case, until the chief pediatric surgeon jumped up and started yelling, "Doctor, you have killed this patient. She was not ready to eat. All her stitches will pull out from the solid food and everything you did will fall apart." He was a conservative old man, and most of the other pediatric surgeons congratulated me for a great job. The little girl left the hospital with a big smile on her face.

Looking back now, I realize that Dr. DeBakey was indeed a great surgeon. He ran a demanding program and didn't tolerate any perceived or real mistakes. He trained a generation of excellent heart surgeons who occupied leading positions all over the world and saved millions of lives. DeBakey's program and discoveries made some of the greatest contributions to medicine.

For me, it had been an honor and distinct privilege to be in his program. Besides surgery, I learned how to be a good and effective physician. I carried with me his drive for perfection throughout my life. One of his favorite sayings was "pay attention to detail". I still follow this in everything I do.

Another great thing about Dr. Cooley and Dr. DeBakey was their ability to attract donations and use them to create the world's largest and best medical center. Instead of just creating a large endowment under their guidance, they used the generous donations to invest in humans by recruiting the best and most promising minds in medicine to the Texas Medical Center. I am eternally grateful for the opportunity I was given to train there.

I still loved my surgery rotations, but I started to get concerned about my persistently abnormal liver test results, which continued a year and a half after my hepatitis. I wondered if I would have the stamina required to be a heart surgeon with my compromised liver. I was also very concerned that I might infect my patients with hepatitis. I went by the pathology labs more often where the now famous pathologist Dr. Harlan Spjut was a senior attending physician. He was brilliant and humble, and he had tremendous integrity. It occurred to me that I might want to get a doctorate in pathology and become a teaching surgeon. It would solve my dilemma of being concerned about infecting my patients with hepatitis and whether I was able to bear the physical demands of a heart surgeon. I discussed this idea with the chairman of the pathology department, and he was enthused about it. We even discussed the subject of my possible thesis. However, I found out that I would have to get a Master of Science degree first, which would add two extra years to my training. That would take too long, so I decided to quit surgery and switch to pathology effective on December 31, 1964,

Pannonhalma's market square with the Benedictine Monastery
and the famous gymnasium on right

Laszlo's Parents' Wedding (October 10, 1925)

Laszlo's Parents' Home with Pannonhalma Monastery in background

Laszlo as a child with his favorite toy about the time he dropped out of Kindergarten

Laszlo and his brother Tibor with their high school caps

Laszlo and his family on his parents' 25th Wedding Anniversary in 1950
from right to left: Tibor (22), Veronica (13) Zsuzsanna (11)
and Laszlo Makk (18)

Medical student Laszlo at Hungarian military training camp in summer of 1954

Laszlo in Niagra Falls in 1957 where he spent the day going back and forth between the U.S. and Canada. Having come from behind the Iron Curtain and its closed borders, something that was inconceivable to him before coming to America.

Dr. Edwards and medical student Laszlo preparing a manuscript for publication at New York State Health Laboratories (1960)

Laszlo's graduation from Albany Medical College (1960)

Laszlo receiving the Outstanding Intern Award in 1961

Laszlo and Carolyn's Wedding in Houston, TX (1962)

Laszlo and Carolyn with Dr. Denton Cooley at their wedding in Houston, TX (1962)

Laszlo in his first car, 1957 MGA

Laszlo's parents arrive at Houston airport from Hungary. From left to right: father Istvan, mother Ilona, wife Carolyn, son Steve, and Laszlo with son Laci (1964)

Laszlo receiving Honorary Mayor of San Antonio, TX award from Major McAllister (1966)

Laszlo's scientific presentation at the International Cancer Congress in Buenos Aires (1978)

The Makk family from left to right: Steve, Andrew, Carolyn, Laszlo, Laz, and Chris (1989)

Laszlo, his wife Carolyn, daughter-in-law Catherine
and son Andrew at the 2006 Kentucky Derby

Laszlo cooking Hungarian goulash at Six Acorn Farm (2009)

The Makks with grandchildren top row: Marshall and Olivia; middle row: Davis, Matthew, Hunter; bottom row: Laci, Sophia and Lucas (Easter 2009)

Chapter XVI

The Making of a Pathologist

I remained in Houston and started my pathology residency on January 1, 1964. Prior to this time, I was able to earn six months of pathology training credit for the electron microscopy work I did while still in medical school, and another six months of credit for my pathology rotation during my surgery residency. Because I had earned these credits, I could compress the usual four-to-five-year pathology specialty training into three years. This also meant I would have to work harder to be prepared for my anatomical and clinical specialty board exams in pathology and laboratory medicine.

Pathology was very easy after the experience I had with my surgery service. I was on call occasionally, but not like I was in surgery when I was on call every second or third night, sometimes going without any sleep, and still having to perform my duties the following day.

Dr. DeBakey had befriended a wealthy Houstonian of Hungarian origin by the name of Ben Taub (the two of us would exchange a few words in Hungarian). On Sunday mornings, Mr. Taub would accompany Dr. DeBakey on rounds with pediatric patients and then have breakfast at the Houston Hilton. These breakfasts led to Mr. Taub generously donating funds to build the Ben Taub Hospital for Indigents (a part of the medical center). Breakfast at the Hilton with Dr. DeBakey was certainly a big deal.

In mid January, I received a call from Dr. Howell, one of Dr. DeBakey's associates. He said, "Laszlo, Dr. DeBakey wants to have you for breakfast on Sunday at the Hilton."

I said that was a big honor and asked, "How come?"

He said, "We just realized we have only three senior residents this July, and we need four. He wants you to come back."

I said it was an honor, but no thank you.

Dr. Howell replied, "Look, you are an excellent surgeon, and you handle the arena atmosphere in surgery so well. You will be bored to death in pathology. We really want you to come back."

I said no thank you, again.

Feeling exonerated for not finishing Dr. DeBakey's rotation, I was very pleased with the choice I had made. I also recalled that when I started my surgical residency, Dr. Jordan said there were 27 residents, but only four would finish four years from now. Indeed, only three remained while the others chose other surgical specialties

For my first rotation in pathology, I was assigned to the VA Hospital. Given it was a 1600-bed facility, a great number of autopsies needed to be performed. As a resident, I was required to also participate in all medical center and medical school conferences, lectures, etc. It was a great learning opportunity. My evenings and most weekends were free. A surgeon from the medical school asked me to teach surgical anatomy to medical students on Saturday mornings. I happily obliged and was happy to receive $10 per lecture.

My training was supposed to be a mixture of anatomical pathology and laboratory medicine, rotating every three months. Once again, I worked with Chief of Pathology Dr. Halpert. Every time I rotated to laboratory medicine, Dr. Halpert's secretary was on the phone within a half-hour or so, saying, "Dr. Halpert wants you to do an autopsy" or "Dr. Halpert wants you to help with a surgical specimen examination." In a one-month blood bank assignment, I got to spend only two days in the actual blood bank, which was getting very frustrating.

My senior resident was Dr. Schmallhorst who had had clinical experience as did I. He was a family doctor before he entered pathology training. Through his connections, we started to moonlight with a pathology service after work for $50 per autopsy. One night, he was performing one of these "moonlighting" autopsies at the hospital in Richmond, Texas when he found a gallstone in a gallbladder that was so large that it completely filled the gallbladder which as a result became distended. Dr. Schmallhorst had the VA photographer take a picture of it, and he put the VA emblem on the picture. Dr. Halpert, who was a gallbladder expert and the author of the gallbladder chapter in the most prestigious pathology book, somehow got a hold of a copy of the photo. Dr. Halpert got very excited about the photograph, and the photographer told him it was Dr. Schmallhorst's case. He immediately called him to his office and wanted to know the patient's identity and why the accession number which links specimen and related materials to a patient's medical record wasn't on the photograph. Dr. Schmallhorst told him there was no accession number, and that the patient's identity could not be revealed. Consequently, Dr. Halpert figured that I probably had the information. He asked both of us into his office and demanded that we tell him the patient's name and hospital number. Dr. Schmallhorst told him that if we revealed the information, there could not be any negative consequences for the two of us. He then told him it was an autopsy finding that he had done while moonlighting. Dr. Halpert almost lost his gentlemanly mind. He told us we had broken the rules by moonlighting, which was forbidden to VA residents, and had used government resources to have the photograph taken. When he rewrote the chapter on gallbladder diseases in the text book *Anderson's Pathology*, Dr. Halpert included the photo and gave full credit to Dr. Schmallhorst.

As time went by, Dr. Schmallhorst became very frustrated about the fact that he was not receiving any laboratory medicine training. He transferred to Hermann Hospital (also a part of the Texas Medical Center) which was the only hospital offering real clinical pathology training at that time. It was a 750-bed hospital: 500 beds for private patients and 250 for the indigent. Luckily, the Hermann Foundation paid for all charity patients. It was an excellent teaching hospital, and the residency was outstanding in both anatomical and clinical pathology. The training at the VA Hospital seemed to place a lot of emphasis on autopsies, while at Hermann they placed more emphasis on surgical and clinical pathology. It was at that time that I decided to transfer my residency to Hermann Hospital. It was a great move, because I received excellent training there. In addition, because Hermann Hospital was within the medical center system, I could participate in conferences and take advantage of other learning opportunities offered by the Baylor College of Medicine and the other affiliated hospitals.

This was a busy place with many lively interactions between clinicians and pathologists. The pathology department also provided services to outlying small hospitals which were not large enough to have a full-time pathologist. Our pathologists didn't particularly like to go there to do the autopsies, so instead of ordering us to perform them, they offered us $50 for each one we did. Actually, that turned out to be a great learning opportunity for independent thinking and thoroughness as well. We didn't have professors to turn to if we had a problem, so we had to solve it on our own.

I used my vacations to do *locum tenens* substituting for pathologists who went on vacation or were sick. At one point, three of us, each with clinical experience, also staffed a 14-doctor medical practice in Houston from 5:00 to 9:00 p.m. on weekdays, and 9:00 a.m. to 9:00 p.m. on weekends. Suddenly, we were living very comfortably with the extra income.

With a third baby on the way, we decided to purchase a house. We found a three-bedroom home with central air conditioning and a fenced-in backyard that was in a good location. The house needed painting inside, and I thought I knew how to paint. I had to paint the living room three times before it was acceptable. One Saturday morning, I was mixing paint in the garage while I was wearing my white intern pants which made me look like a painter. A neighbor came over and said that he understood a doctor was moving in. He wanted to know what the doctor was like and what his name was. I told him that the doctor was an okay guy and that his name was Laszlo Makk. After some small talk, he introduced himself as he was leaving. I then told him my name, and he asked incredulously if I was the doctor who was moving in. When I said yes, he was very embarrassed.

Shortly after we moved in, I was told that the divorced husband of one of my neighbors had taken off from the air strip on their ranch in his plane and had disappeared two weeks ago. A few days after hearing this, Dr. Wilson Brown, Chief of Pathology at Hermann Hospital, called me around midnight and asked if I would do an autopsy on a terribly decomposed body. He said he would assist me. It was to take place in the basement of a funeral home. The electricity had been knocked out earlier in the evening during a storm, and it was not back up yet. The heat and the smell were nearly unbearable. The body bag beside the decomposed body was full of maggots and other creepy crawling bugs that were running up my arms. The body had been found hanging from a tree near an airplane wreck and was burned beyond recognition. Even though his wallet had been found, I took his jaw for dental identification. Indeed, it was my neighbor's ex-husband. The newspapers were full of the story. I found out that he had remarried, and that this younger woman had taken out a considerable amount of life insurance on her old husband. These circumstances called for a thorough investigation of the crash.

Another interesting case in the hospital came along, and I decided to prepare it for publication. The article was published in a local medical journal and was well received. Dr. Brown suggested that I prepare and submit a surgical pathology case report every month for publication. This involved case selection and the challenge of working within a three-page limit to include all text, gross pictures and microscopic photographs, regardless of how complex the case was or how large the medical record). I also had to review the medical literature on each of these cases. For two years in a row, I received the Violet Keeler Award for these publications which provided me with a $100 stipend monthly over and above my regular salary. That was big money at that time.

I was busy with my residency and new life in America, but often thought of my family in Hungary. Since escaping from Hungary, I had only seen my parents twice. There was one time when my mother was able to get out of Hungary and travel to Vienna where my brother Tibor lived. It would be a great opportunity to visit her since Hungary's borders were closed. But, I just didn't have that kind of money. I had to give up on that until Air France called me and said, "Dr. Makk, we have a round-trip ticket for you to go to Vienna. Where do you want us to deliver it to?"

I told them it had to be some kind of mistake because I had not ordered or paid for a ticket. I asked "Who ordered the ticket?" But, he said he couldn't reveal the identity of the person.

The visit was terrific. I found out later that the generous mysterious and generous patron was John Bender, my dear friend from Albany, New York who was like family. When he heard about my problem affording the air fare to visit my mother, he bought the ticket, sent it anonymously, and wouldn't let me pay him back for it under any circumstances.

(Years later after completing my residency and going into practice, I did make it up to him. I had to be in New York City to present a scientific study in laboratory automation. This coincided with Mr. Bender's birthday, so for a gift, I gave him a penny minted in the year he was born and wrapped it in a 1000-dollar bill. I told him I had a special birthday present for him, but before I would give it to him, he had to promise me that he would keep the wrapper. He got the biggest kick out of that. It was just a small token of my appreciation for the kindness, support, and love the Benders had enriched our lives with over the years. Even such a present given each day would not fully express our gratitude to them.)

After visiting Vienna, not much time passed before I learned my mother had become ill. She had a bad heart in addition to needing two operations to survive. I feared that if doctors had put her to sleep in Hungary, she might not have awakened, so we decided to bring my parents to America. Hungary was still under Communist rule with closed borders, but because she was sick and my parents were old, the government was willing to let them leave peacefully.

It was in 1964 that my parents arrived in the United States for the very first time. Of course, it was a very emotional occasion. Eight years before, I had been a penniless refugee, and my parents didn't know if I was dead or alive. Now they were in America to see their son, an American citizen who was not only a doctor in the U.S. working in a world-famous medical center, but the husband of a beautiful American wife and the father of two fine boys.

Because 1964 was an election year, campaign signs could be found everywhere. As we drove home from the airport, my father pointed to a Goldwater sign and said, "He is a good man. Communists hate him."

Dr. Sam Law, the brilliant, young chief surgeon at the VA Hospital, was now in private practice and operated on my mother. It went well. During her recovery, my parents learned that I was working for the Goldwater campaign. On election night, we had a little party. As the numbers came in and showed that Goldwater was losing, my father looked sick. I asked him if he was okay. He said, "Yes, but what about you?"

I asked him what he meant by that.

He said, "Your man is losing. What is going to happen to you? Are they going to arrest you?"

I said no.

Then he asked if I would be able to keep my job as a doctor. I told him that my job was safe. He said, "Thank God!" This was how terror and fear under Communist rule affected people's thinking.

We were hoping my parents would stay with us instead of returning to Hungary, but they said they were too old to learn English and get jobs. They didn't want to be a burden on us, and my father wanted to die in the house he had built.

Our life was happy, and with all the moonlighting jobs, we were free of the usual financial worries of most residents. It turned out that our mortgage payments were not much different from our rent. The Ford station wagon that we had purchased several years before for $400 was about to fall apart, so I started looking for another car. I saw an ad in the paper for a Mercedes 220S at a suspiciously low price. Black with a beautiful red interior, it was three or four years old with only 50,000 miles on it. The owner was a sea captain who was buying cars like this in Europe and bringing them back to the United States on his freighter. He was ready to sail again, and I knew he wanted to sell it before he left. I told him I didn't have enough money. All that I had was $2,500, and if he would let me have it, I would pay cash right away. He agreed. It was unbelievable. The car was so nice that it stood out even in the doctors' parking lot.

Around that time, Jones Hall, a new performing arts center, was being built in downtown Houston. We decided to go to the opening night which featured an opera. How often do you actually have the opportunity to be in the audience for something like this? I polished the Mercedes twice. In the meantime, I acquired a chauffeur's cap and white gloves. I asked Carolyn to sit in the backseat so we could accommodate another couple who was coming with us. After we got on the road, I put on my chauffeur's cap and white gloves as a practical joke. When we arrived, it was impossible to find parking, so looking the part with my chauffeur's cap and shiny Mercedes, I just eased myself into the line of limousines being directed to the VIP parking adjacent to the concert hall. The police just moved us on. Once again, I learned there is no penalty for trying!

After the opera opening, I was back to work and became Chief Resident. I got to do all of the diagnostic frozen sections from surgery. When I needed help with a diagnosis, Dr. Brown would say, "Figure it out, and you'd better be right, because as a pathologist, you get paid to be right." (Then he or another pathologist would step in and help.)

As I was getting close to finishing my resident training, I received quite a few job offers, and some sounded quite lucrative. I attributed this to my publications and *locum tenens* work.

One day, Dr. Brown called me to his office, complimented me for my work and drive, and said, "A year from now, you will be a chief of pathology and laboratory director somewhere because you have what it takes!"

I was elated and thanked him right away.

Near the end of my training, the department sent me to the College of American Pathologists Convention workshop to learn new laboratory procedures. Only two of us were pathologists among the participants, and we ended up working together at the convention. The other pathologist was Dr. Ed Fadell from Louisville, Kentucky. After a while, he became very inquisitive about my background. We had lunch together, and as we parted, he wanted to know if it would be okay for him to give my name to a hospital in Louisville which was old, but doubling in size and would be looking for a new chief pathologist. I agreed and thanked him for the thoughtful consideration.

Of all the offers, I chose the one that paid the least but was the most prestigious and showed the most potential for furthering my knowledge. In most medical schools, a new doctor usually starts out as an instructor for three to four years. Then, if the person is performing well, he or she may be promoted to assistant professor. Baylor College of Medicine offered me an assistant professorship coupled with a surgical pathologist position at Methodist Hospital which by this time was the crown jewel of the Texas Medical Center. All of my colleagues were excellent medical professionals. I felt highly honored to work with them, and especially with the illustrious Dr. Harlan Spjut—my brilliant hero in pathology.

Before I took this position, I was invited to several places for job interviews. Carolyn accompanied me. The first was in a small historic town in Virginia. Everything went well with the very cultured chief pathologist on Saturday. On Sunday, I went to see his associate. Instead of diplomas hanging on his office walls, there were large pictures of Jesus and Robert E. Lee. He let me know he was a proud Virginian and had only left Albermarle County when he was "conscripted" to the U.S. Air Force. He didn't hide his prejudices and wheedled into the conversation several times that my accent would probably be okay. Finally, I had had enough of this and him. I told him that I spoke four other languages beside English, and if he had a problem with my English accent, we could try the others. I thanked him for his time and left. Life was too short to associate with jerks like him.

We had the afternoon free; however, there was little to do. The motel was out of newspapers, and they didn't even have any books to read. The TV didn't work, and the only other entertainment was a quarter-operated vibrating bed. Out of boredom, I started to read the yellow pages. When I saw that there were twice as many preachers as doctors listed, that was it. We were leaving.

Prior to this, I had been contacted by the hospital administrator in Louisville to whom Dr. Fadell had recommended me. She did invite me for an interview, but I had left it up in the air. Because we ended up leaving Virginia a day early and were close to Kentucky, I thought it might be a good time to fly to Louisville and check out the hospital. St. Anthony Hospital Administrator Sister Francis Ann, a Franciscan Sister, was glad to hear from me when I called. I asked for the name and phone number of the current pathologist, following the physician code of ethics to always contact an incumbent physician first (and not go behind his or her back).

I called Dr. Cummings and told him I was finishing my training, and that I understood the hospital was doubling in size. I thought he might have a need for an associate in the pathology department, but he said no. I asked him if I could buy him dinner, lunch, or breakfast, or if I could just come by to introduce myself. Again, he said no. I told him that the hospital administration and medical staff were looking for a pathologist, and that I had been invited for an interview, but I wanted to talk to him first as protocol demands. If he wouldn't see me, I told him I felt I must cancel my interview and leave.

Dr. Cummings finally invited me to come by his office and asked that I be there at nine o'clock the next morning. We arrived in Louisville the night before. Everything smelled like sour mash bourbon whiskey. The people were very friendly, and the Brown Hotel was very nice.

I arrived at the hospital promptly at nine o'clock the next morning; however, Dr. Cummings made me wait an hour before he saw me, and then he didn't even ask me to sit down. He did ask me if I had any medical publications and where they had been published. I told him I had published 16 articles. I didn't know at that time that he was not published nor was he board certified. He told me again that he didn't need an associate, even though he said he was aware they were looking for a new pathologist. It was okay with him if I talked to the medical staff and hospital administrator and that is just what I did.

My first meeting was with Sister Francis Ann, the administrator. The second was with a bright young family doctor by the name of Dr. Vanderhaar, chair of the medical staff's executive committee, who later became the first Chairman of the Family Practice Department at the University of Louisville School of Medicine.

The interviews went well. When they asked me what it would take to get me to take the position, I presented a list of recommendations from the College of American Pathologists' contract manual. The manual recommended that the pathologist be an independent contractor rather than an employee, be able to hire and fire all personnel under his direction, have financial control over the department, have all rights and privileges of other medical staff members, could be dismissed only for cause by the hospital, and that his compensation should be a percentage of the departmental income. I realized this was a first meeting and didn't expect a firm offer at that time. I wasn't aware that there were 37 other applicants seeking the position.

Finished with my meetings, Carolyn and I drove around the countryside, and it turned out that I loved Kentucky. Carolyn was originally from Kentucky, and when I asked her if she would like to live there, she just kept saying, "I'll be happy wherever you want to practice and choose to live." We flew back to Houston shortly after that.

Some time later while I was out duck hunting, Dr. DeHaven (the incumbent pathologist from St. Anthony's Hospital) called, and asked Carolyn what job offer I was going to accept. She told him I was going to take the position at Baylor/Texas Medical Center. He just hung up.

A few months later, Sister Francis Ann called and offered me the position of Chief Pathologist and Director of Laboratories at St. Anthony Hospital, and asked me to come to Louisville to work out the final details of my contract. I asked her to provide the airplane tickets this time, a request to which she readily agreed. I also asked her the name of the hospital's current pathologist. She told me that Dr. DeHaven was leaving and that a Dr. Allen was helping out during the transition. I got in touch with both. I asked Dr. DeHaven if it was okay with him that I come to Louisville for contract talks, and he said of course. I asked him to tell me why he was leaving. He said, "It is a good job, and maybe I should kick myself in the butt for leaving, but I just can't stand these holy statues around here."

I called Dr. Allen who verified for me that he was just temporarily helping out while a department head was being found. He said he would even pick me up at the airport, and I would be able to recognize him as the short fellow with a big cigar.

I arrived in Louisville at 7:00 p.m. on a Saturday for a meeting scheduled with the Sisters at 8:00 p.m. I was going to stay in the hospital guest suite, but Dr. Allen insisted we stop by his house to meet his family and have a quick drink with them. I had not had anything alcoholic to drink since my hepatitis had threatened my life, and after the first drink, I was already high. The talking was intense, and I couldn't even get to a phone to call the hospital to let them know I had been delayed until 8:30 p.m. Sister Francis Ann said that if I could get there before 10 o'clock, we could still have our meeting. Otherwise she would see me at eight o'clock the next morning. I didn't get in until 1:00 a.m.

The next morning, I woke up with a big hangover. This was embarrassing, and it was the last thing I ever wanted to happen to me, particularly on that morning. She greeted me by saying, "Doctor, you are looking pretty good with the little sleep you got." Oh, big sister was watching!

I brought my contract proposal which was based on the College of American Pathologists contract manual. Experienced senior colleagues and legal counsels had written it. The meeting was business like with no major disagreements. I left a copy of my contract proposal with her, and she said we would meet again the next day, when the attorney for the order of Franciscan Sisters would be there from Chicago.

The next day, the attorney tried to tear apart every major issue in the contract. Pretty soon, I realized that if I gave in, I would be an indentured servant rather than an independent contractor with limited authority while being held totally responsible for lab operations. He argued on every point of my proposal, and the meeting extended into the evening. He accused me of being argumentative, and said that I talked too much. I raised my voice somewhat and told him that every part of the contract proposal was meant to reassure the creation of an excellent laboratory and pathology service to the patients. If he didn't like my personality, I could find a really quiet, non-troublemaking pathologist who would be like an umbrella that he could bring in with him in the morning, put in the corner, and take home in the evening.

The lawyer's attack-style tactics evoked a lot of sympathy for me from the Sisters. Of course, I had a couple of points in my contract proposal that I had put in to "give in on" during negotiations. After I found out that the incumbent pathologist had flunked his specialty board examination three times, I added to my contract that the contract would remain valid only if I passed the medical specialty board exams on clinical and anatomical pathology, which I was planning to take anyway.

Because I had a good position in Houston, I decided we had negotiated long enough. I politely told them that negotiations were over, and they could either accept my contract as is or not. In a short while, they said the contract was accepted. I couldn't believe it. I had obtained everything needed to build and operate a good laboratory.

They offered to type the final form of the contract, and we would sign it the next morning, so I booked a flight back to Houston. The next morning, the contract was not finished. I had a taxi waiting, because I needed to get to the airport as fast as possible. Finally, the contract was ready, but I had a copy of the draft, and instead of just signing it, I stopped to compare the draft with the final copy. I was glad I had, because "inadvertently," the paragraphs on my compensation and my authority over lab personnel had not been included. In the big hurry, the paragraphs had supposedly been omitted. We then added these to the contract, and then I signed it.

I gave a big tip to the taxi driver to get me to the airport as fast as possible because I was about to miss my flight, but we made it just barely. As soon as I boarded, we took off. I thanked God for His help negotiating a fantastic contract. Before I left Houston, I would have been happy to get half of what was in the draft, but instead I had it all. I now had my chance to build a good laboratory.

After my return to Houston, I had to get ready to take my clinical and anatomical pathology board exams at the same time while still handling my regular job responsibilities. I found out that about 70 percent flunked clinical pathology. and that 20 percent flunked anatomical pathology. Passing my boards was a sensitive issue since, out of goodwill, I stipulated in my contract with St. Anthony Hospital that I would be board certified in both clinical (sometimes referred to as laboratory medicine) and anatomical pathology.

One day, I got a letter addressed to Laszlo Makk from Dr. John Anduhar, President of The American Board of Pathology. The letter made reference to the article on comparative electron microscopic study I had published with Dr. Edwards while I was still a medical student. He asked to borrow the negatives of some of our electron photomicrographs. What luck! I found them in the attic and sent them with a cover letter.

Dr. Anduhar served as proctor during my board exams which were held in Miami. While he supervised, he stopped by my table and thanked me for sending the slides. He said they were going to use them in the exams, but when they saw my name among the examinees, they decided not to use them. He also complimented me on the quality of the pictures.

The three long days of examinations were extremely tiring, particularly in clinical pathology, where our sufficiency in clinical chemistry, hematology, blood banking, bacteriology, parasitology, and mycology was tested. In parasitology, we were shown a lot of pictures, which looked like parasites but were not. This included eight or 10 cases in sequence, where the correct answer was "none of the above." After the exam, there was an examinee who was bragging about how he had known the answers. I told him about my "none of the above answers," but he said they were all parasites. That worried me until the post-exam conference when we were told the correct answers were "none of the above," because they wanted to make sure we would not call parasite-looking artifacts parasites and subject the patients to unnecessary treatment.

Finally, we finished the exams, dead tired. My mother-in-law who lived in Florida came to meet me. We had a couple of martinis in the hotel bar. As I went back to my room, the elevator was full of my colleagues. At one floor, Dr. Anduhar stepped into the elevator and asked how the exam had gone. Then he looked at me and said, "Isn't it nice, Dr. Makk, that you only had to take them once?" My head was a little hazy from the martinis, because

I hadn't been drinking since my hepatitis, except for the one time at Dr. Allen's. I didn't realize that he had just told me I had passed. There was a rumor that if one ran into a board member at the airport, they had a printout of the exam results, and they would usually tell you if you had passed. I spent the next day at the Miami Airport, hoping to run into an examiner, to no avail. I needed to know the results, because I had made board certification a condition of my contract with St. Anthony's Hospital.

After six weeks of waiting, the letter finally arrived, notifying me that I had indeed passed both the anatomical pathology and clinical pathology board exams!

In the meantime, one of the senior surgeons, Dr. John Overstreet, came to my office and asked, "Laszlo, what do I hear that you are moving to Louisville, Kentucky?" I asked where he had heard this, and he said a Louisville cardiovascular surgeon whom they had trained had called and wanted to know if I was any good. Jokingly, I said that my leaving depended on what he had told them. Dr. Overstreet assured him that I was a good doctor.

After my board certification letter arrived, I submitted my resignation with an effective date of when a suitable replacement could be found. The hospital administrator told me I was in line to one day be the head of the pathology department and director of laboratories. I was highly complimented.

In the meantime, I heard a Baylor-trained pathologist was just getting out of the military. I contacted him and asked if he would like to have my job. At first, he thought I was kidding, because Methodist Hospital was the most prestigious institution at the world-renowned Texas Medical Center. When I told him my situation, he immediately applied and was accepted.

I was on my way to Louisville now! But first we had to sell our house. We started out without a realtor in order to save the commission, and a "for sale by owner" sign was posted in our front yard. Our location was excellent, and after less than a week, a family from Albany, New York, drove up and took a good look around. They said they would take it, offering to pay the asking price, and would be back the next day with the contract and a cashier's check. They didn't show up the next day, and we started to get nervous. Finally, they came three days later with the contract and check. We realized a nice profit after we had owned the house for four years. By the time we had completed the paperwork, I was able to find a person who would drive my MG to Louisville. This way, the family didn't have to travel in two cars. We just all went in the Mercedes.

Chapter XVII

St. Anthony's Hospital and Kentucky

The trip to Louisville was full of hope and happiness. For Carolyn, it was somewhat of a homecoming, because she was born in Kentucky. We also experienced wonderful Kentucky hospitality in her hometown of Smith's Grove where a cousin welcomed us with a wonderful lunch in her beautiful home with large white columns out front. After lunch, we visited the cemetery. Then we drove on to Louisville, where I had leased a house during my last visit. The moving truck was supposed to arrive at the same time as we were, but it wasn't there. This was before the cell phone, and after we waited around in the empty house, we checked into a nearby motel where we had to smuggle in our dog.

I called the moving company, and they informed me that the front wheel of the tractor had broken, and that it would be on the way after repairs but might take some time.

During our first night in the motel, Laci, our eldest son who was five at the time, had such a severe asthma attack that he almost choked. I left to look for a drugstore, but they were all closed. Finally, around 11:00 p.m., it hit me: the hospital should have a pharmacy that would be accessible at night. Only the night nursing supervisor had the key to the pharmacy, and she didn't know what to do with me. She was faced with a guy who had a strong foreign accent and who called himself a doctor and wanted to get into the pharmacy at night. Finally, I prevailed and got into the pharmacy and found the asthma medication. When I got back to the motel, Laci was still choking; however, the medication worked quickly, and he went to sleep shortly thereafter.

The next day, I went to the hospital and introduced myself to the laboratory personnel and interviewed them to get an idea about their skill levels. Some of them had no formal laboratory training, and the quality of their performances was questionable. Quality control was an essential part of reliable and accurate laboratory testing. For example, if we perform a test to determine a patient's blood sugar level, we first need to run sugar tests in high, low, and normal levels to make sure that the results would be accurate in all levels.

When I asked to see the quality control documentation, I was told that they traditionally sent those to the Baptist Hospital across the street. Later, I also found out that many of the routine tests that could be performed in the lab were also sent over the Baptist Hospital. Fortunately, we had a couple of registered laboratory technologists with whom I could start to build a good quality laboratory.

The situation was even worse with autopsies. I found that in the past five years, only one was reported, and the only documentation on dozens of cases was the patient's name in the autopsy accession book. The specimens were not preserved, nor were microscopic examinations done. With my clinical experience coming in handy, I reviewed all of these cases, and instead of autopsy diagnosis, I signed them out as pathologist diagnosis based on available medical record review.

Next, I called on the chairman of the medical school pathology department and on all chief pathologists in town to introduce myself, which was appreciated very much.

The mover still hadn't arrived, so we continued to camp out in the motel. One evening, Dr. Herbert Dickstein knocked on the door. He was a well-trained pathologist from New York, and his wife had been accepted to medical school in Louisville. They moved here, but because his first job hadn't worked out, he was looking for a position that was available right away. After a brief reference check, I decided to take him on, even though it meant I would have to share my compensation, which reduced my earnings considerably. It was the right decision because now I had time not only to do my routine work, but also to develop the laboratory and supervise its construction which had just started.

This turned out to be more time consuming than I had expected. I traveled to the best places to learn about laboratory design and automation. Our architect was kind enough to let me make seventeen changes to the original design. At that time, laboratory computers were in their infancy. After I visited the most computerized laboratory, I concluded that computerization added only to the operational expenses at that time and wouldn't add anything to accuracy, precision, or speed of reporting. Ours eventually became the most automated and fastest laboratory in the community. Instead, we implemented lots of new laboratory procedures, many of which had been sent out to other hospitals, and at that point, they started sending us tests that they didn't perform.

I wrote a laboratory manual for our medical staff. I did best writing it in the morning between 4:00 and 7:00 a.m. when it was quiet. Fortunately, Dr. Blaine Lewis, a surgeon, had a hobby print shop and printed it for us. When we set up our first multiphasic auto-analyzer, which performed 12 tests most likely to indicate disease, we studied the results of 1000 patients. We determined whether these tests confirmed the clinical diagnosis, suggested diseases that were not contemplated by clinicians, uncovered any disease in an early stage before clinical signs of it developed, or if they just confused the clinician. I presented our findings at the International Congress on Laboratory Automation, and my paper was printed in a book.

We were on our way. As we put in new procedures, I kept our medical staff informed about their clinical usefulness in newsletters. We participated in the College of American Pathologists Laboratory Inspection and Quality Improvement Program, which among other things, consisted of sending samples of tests to us. We compared the test results to national standards.

In anatomical pathology, I made a rule that any surgical specimen removed from patients by 5:00 p.m. would have a pathology examination report the next day, preferably by noon. Necropsy reports were to be completed preferably three days after death instead of months afterward so that the death certificate would also be based on postmortem findings. We also put one of us on call around the clock for frozen sections and other laboratory consultations. I situated the morgue so that it became part of the laboratory in an isolated area, which proved very handy for the doctors attending their patient's postmortem examination. We also made attending autopsies available to nurses and students of all kinds. I had an X-ray viewer installed so we could review X-ray findings with radiologists and compare them to autopsy findings. At times, the autopsies became almost like medical conferences with clinicians and nurses present.

The availability of a frozen-section consultation at anytime was appreciated by the medical staff. Frozen sections provided immediate diagnosis. They were particularly useful with breast biopsies, because if the diagnosis was cancer, the surgeon could proceed with further surgery right away. Before, he would perform the biopsy and do the definitive surgery after the biopsy was diagnosed at a later time. The new way spared the patient from a second operation and second anesthesia, and it shortened the hospital stay. This was also done on uterine cervical cone biopsies and other removed diseased tissue.

Our major strength came from continuous efforts to educate our staff and ourselves. We put the patient in the center of our activities. We even went as far as coordinating blood drawing from hospital patients in the morning before coffee or breakfast was served.

Getting registered medical technologists was very difficult, because medical laboratory procedures were expanding and there were simply not enough of them to go around. I decided the only way to remedy the situation was to start a medical technology school. To be accredited, we had to institute a well-structured school and had to pass a tough inspection.

In our school, students had to have at least three years of college with a science major and then spend the fourth year in our program. On successful completion, students would receive a bachelor's degree from their parent college and a certification from us in order to take the Medical Technology Registry Examination. On passing this exam, the candidate became a registered technologist, BS, and the best-qualified laboratory professional. We took teaching very seriously. It also helped our employees become more skilled. Eventually, we became affiliated with 10 Kentucky colleges that could send their students to our school.

Our School of Medical Technology, along with other schools, eliminated the registered medical technologist shortage. Our emphasis for continuous learning sunk in because many of our students attained senior positions in other laboratories. Quite a few obtained medical degrees, and some of them became pathologists. As I look back, the school and emphasis on continuous learning among the laboratory staff, giving everybody a chance to improve themselves and work their way up, was one of the best things I ever did. We had people who started out as filing clerks and then became phlebotomists, nurses, or technologists. New laboratory testing methods became available at such a fast rate that it was either keep up or you would be left behind.

The school gave prestige to our department at St. Anthony's, and we no longer had a shortage of technologists. When we had an opening, many of our graduates would apply for it and come back to work, sometimes with a pay cut.

We operated the School of Medical Technology for nearly two decades. As the era for laboratory automation started to set in, the new instruments replaced many of the jobs in medical laboratories, and a shortage of medical technologists turned into a surplus. I didn't see any reason to take college students with 3.6 or 3.8 GPA and make them study and work hard to earn a bachelor's degree, only to find a job market so limited that some of them could only get a part-time job on a night shift. With that in mind, I suspended our school for a few years and then closed it.

Finally, everything at St. Anthony's was organized, and it was time for my family to take a vacation. In 1969, we decided to go to Europe. My parents were able to get out of Hungary and go to Vienna to visit my brother's family, so we decided to join them. We figured if we were going so far, we might as well tour other parts of Europe as well.

Our children were seven, six, and four, and the baby, Andrew, was just five months old. Carolyn and I needed to apply our organizing skills for the arrangements. For instance, we watched how many diapers Andrew needed in a day, and by multiplying that by 30, we had the number of diapers we would need to take with us. After we factored in two diarrhea episodes, we ended up with a large suitcase full of disposable diapers. We added a self-sealing plastic baggie for each diaper for disposing purposes. This didn't work out very smoothly because most of the hotel help was not familiar with disposable baggies. Housekeeping took the dirty diapers out and washed the baggie for us to put back with our belongings.

We thought we were ready and organized. Everything had been coordinated with the European relatives. Then the day before we were departing, Andrew came down with chicken pox, and to go or not to go was the big question. Eventually, we decided to go on the trip. We landed in London the next morning and tried to figure out what we would do if the customs people wouldn't allow our five-month-old baby into the country because he had a contagious disease. By the time we landed, his little body was covered with red papules. Thankfully, he was asleep when we went through customs, and the official barely looked when I opened his blanket a bit.

We were supposed to have two adjoining rooms everywhere we stayed. However, the London hotel gave us one room on the second floor and one on the fifth floor. The hotel was fully booked, and they couldn't rearrange the rooms. With Carolyn's great concern, we decided to let the three boys stay in one room while Carolyn and I stayed in the other. We instructed the boys not to open the door under any circumstances. The boys did just fine.

Our next stop was Switzerland, where we had a wonderful stay in a formal place in Lucerne at the Chateau Gutsch. One night, as we were finishing dinner, an elegantly dressed lady approached us and asked if we were British, because our children were so well behaved and had the nicest table manners. I proudly told her, "No, ma'am, we are Americans."

We had a driver with a Volkswagen bus drive us through the Alps to Italy. The first stop was in romantic Lake Como. I fantasized about this place ever since I took high school geography, and now, here I was! We stayed in an elegant villa that had been converted to a small hotel near the lake.

There were little bicycle boats on the shore. I could hardly wait to get into one with my three older boys. After we got into the boat, a strong drift started to pull us toward the center of the lake, where a big steamer was creating big waves. I kept waving for help, but people must have thought I was just having a good time and waved back. Afterward, my legs were very sore from pedaling so hard. Carolyn had stayed on the balcony with baby Andrew to watch. I thought she would see us and help. I kept waving and screaming for help, but she thought I was just waving hello. I could not get to the shore, but after trying harder, I made it to shore and just laid there and didn't move.

After I recovered, we dressed for dinner. I told the waiter to bring the specialty of the house. It was brought to the table by two waiters in a large silver dish with a very large silver cover. When I saw this, I started to salivate like Pavlov's dog did when he heard the bell. After the cover was removed with great pomp, there was just a bunch of asparagus with a fried egg on top. Carolyn and the kids had a delicious dinner, but I was still hungry; however, the only thing I could do was pay the check.

The Lake Como experience wasn't over yet. Carolyn and I had a large bedroom with a marble floor, two single beds adjoining, and the kids were next door. After they went to sleep, I was easing over to Carolyn's bed when the two beds separated and I fell to the marble floor and hit my head so hard on the that I thought I had a skull fracture. That woke up the kids, and I was ready to get out of Lake Como.

Then I learned about Italian drivers. Assisi and Rome were great, except for one incident. As we were boarding the airplane, a policeman stepped in my way and demanded forty American dollars to let the baby board. It was outright blackmail, but if we didn't pay, we would miss our flight. Finally, we settled on $20.

Vienna was great, and the family reunion was overwhelming. However, on the way back, we all got intestinal flu. Luckily, we made it back home safely. I learned from this trip that Europe was nice, but America was the greatest. Now it was time to get back to work.

Chapter XVIII

Storm Clouds in Paradise

Things at St. Anthony's were progressing well. In addition to offering reliable laboratory services, the radiology department was upgraded with new, young, talented radiologists and many new procedures. A brilliant new radiation oncologist and radiation treatment center was also added to the hospital. We were quickly becoming a cutting-edge medical center. Sister Modestina retired as head of the business office. One day, Sister Francis Anne, the administrator, came by my office with a well-dressed young man and introduced him as the new Director of Finance who was replacing Sister Modestina. Mr. Hall was pleasant, and we both expressed the desire to work with one another. He seemed familiar to me. He told me we had met before when he had audited the hospital for Medicare.

Shortly after I had arrived at St. Anthony's, he had come by the lab as a Medicare auditor and had requested that I show him 15 Medicare charge tickets. I picked up a batch of charge tickets and separated those ages 65 or older. He looked at me and said in a stern voice, "You don't even have them separated for Medicare and non-Medicare?"

I told him that I didn't.

Then he grinned and told me that that was the correct answer, because if I had them separated, I would have been breaking the law. I then thought, *Why is this little pipsqueak trying to set me up like this?*

Initially, we got along. I was glad to see a hardworking young man with a business sense on board. Then things started to change. He started to hire all kinds of directors and managers to report to him, but none of them had anything to do with patient care.

One day, Sister Modestina came by and asked me to be her doctor. I explained to her as delicately as I could that, being a pathologist, none of my patients were living. I referred her to a good doctor; however, she immediately said she had seen them all in her 60 years of hospital service, and she still wanted me to be her doctor. She was in a very chatty mood and kept on talking. She said that when she turned 84, she was told she had been getting slow, and that they had replaced her with Mr. Hall and eight executives in the business office. Then she alluded that no one seemed to know anything about patient care or hospital business anymore. After she left, I kept thinking that she was right. Shortly thereafter, she was removed from the hospital.

We had a very bright Assistant Administrator named Ben Hessen, who was above Mr. Hall. One day, he disappeared from the hospital, even though he was capable and liked by many. All kinds of rumors circulated about his firing, and all seems to point to Mr. Hall. The Director of Maintenance told me he was fired on the spot without any reason because of Mr. Hall's instigation, and he was not even allowed to clear his desk. I was curious about his shocking removal, and I called Mr. Hessen at home to find out what had happened. He said he had been talking on the phone in his office and had heard some clicking. He checked the receiver, and there was a listening bug in it. He left it untouched, locked his office, and went to report it to the police. When he returned, his office was open, and there was a letter on his desk informing him of his immediate dismissal. He could not even find out why. This disturbed me, because he had four young children, and this was terribly destructive to his career. The next day, a decorator came to redo Mr. Hessen's office for Mr. Hall. After that, it seemed a high morale in the hospital had been replaced by fear.

A while later, I got a call from one of our senior medical staff members. He asked if he could give my name to the board chairman from an Indiana hospital that was trying to get a recommendation on Mr. Hessen. No one from the hospital administration was answering his phones or returning his repeated calls. I told him I would be glad to recommend him. When the board chairman from Indiana called, I told the chairman that Mr. Hessen would do a good job in that capacity, and he was hired.

Shortly thereafter, Sister Francis Anne asked me to stop by her office. She seemed like a different nun now. Her kind and friendly manner changed to cool and scolding. She told me I had made a mistake by giving Mr. Hessen a good recommendation because he was not qualified for the job. She said my recommendation would reflect badly on our hospital. I told her I had given my recommendation as an individual, not as a hospital official. It seemed like someone had programmed her. When I asked why Mr. Hessen had been fired, she did not answer. Then I wondered how she had found out that I had given him a recommendation. I did not have to wait too long for the answer.

One morning, the FBI called me at home and requested a meeting. Two agents arrived 10 minutes later. They informed me that my phone had been bugged in the hospital and that they needed my cooperation, because it is a federal crime. They told me someone from the hospital had purchased several illegal phone-bugging devices and had planted one in my phone. I told them that about a month before, I had been on the phone with a friend when my phone started to click. He told me the sound was likely the tape running out on a listening device. At the time, I thought he was kidding, but he was dead serious.

Now it became clear that Mr. Hall had taped my phone conversation with the Indiana hospital board chairman, twisted the information, and passed it on to Sister Francis Anne. It now seemed that Sister Francis Anne was under his spell, too. I discussed this incident with my attorney, and he assured me that I didn't have to worry. He said that it would be very costly to them if they touched me, and that I had no reason to worry.

We opened a private laboratory outside of the hospital at the request of medical staff members so that it would easier for their patients to obtain lab work. Our private laboratory had an arrangement with the hospital. We would send tests, which were performed on a high-capacity automated analyzer, to the hospital lab, because their equipment was only utilized a fraction of its capacity. This arrangement benefited the hospital in two ways: First, the more tests run on their instruments, the less expensive the unit cost, and therefore, cost per test was lessened. Second, the hospital received revenues that they otherwise would not have. Our lab paid for every test, and our lab benefited as well, because we did not have to invest in expensive testing equipment.

Most hospitals were grateful for such arrangements. The first of every month, I presented a check to the hospital for the previous month's work with a note on the check that said "for referred lab work". We even paid the hospital for tests that we did not bill for; such as clergy, doctors' families, and nurses. One day, Mr. Hall had asked me if I would mind not writing "for referred lab work" on the checks, because he could better utilize the money in the hospital's general fund.

It was fine with me until a few months later when our medical staff representative to the hospital board stopped me, and said that I had been discussed quite a bit at the last board meeting by Mr. Hall. He told me Mr. Hall said I was ripping off the hospital by having all kinds of tests done for my private lab and not paying for them. He said I only occasionally gave very modest contributions to the hospital. He also told me that he was trying to get me fired. I asked him if he had explained that our hospital lab was consistently being voted the best in Louisville. He said it was a busy agenda, and he didn't want to take up time with his comments. I did not have to wait too long to hear something about it.

About a week later, I was doing an autopsy, and my secretary came into the morgue and said that Sister Francis Anne had asked if I could come to her office right away. I rushed down there, still wearing my scrub suit. The Board Chairman, Administrator Sister Francis Anne, and Mr. Hall were all there. The chairman told me that they didn't want me to refer any more lab work to the hospital anymore. I asked him why, and he referred the question to Mr. Hall who said something to the effect that Medicare may bring criminal charges against the hospital if they accepted lab referrals from me.

This was an obvious lie. Mr. Hall then asked if I would turn over all records from our lab to the hospital so they could audit them and make sure my payments were correct. I told them our records would not leave the lab, but that they were welcome to audit them there on site. It was then that I realized I was Mr. Hall's new target; most likely because of my support of Mr. Hessen or because I didn't give up my independence to this self-proclaimed despot.

The Motherhouse in Mishawaka, Indiana, was the headquarters of the Franciscan Sisters that ran St. Anthony's Hospital. They supervised their 11 hospitals from that location. Mr. Hall informed me that the Motherhouse had hired an attorney to handle the audit. The lawyer was an uncooperative "SOB" who didn't even return my attorney's phone calls. My attorney felt very strongly about not turning over our records to the hospital, because under Mr. Hall's "audit," they would accuse me of something I was innocent.

Once, I was in Mr. Hall's office discussing laboratory business matters when the conversation suddenly turned to guns. He opened the center drawer of his desk and pulled out a fully loaded 9mm gun. When I asked him why he kept a gun like that in his desk, he said, "Any SOB who would want to hurt me won't leave my office alive." I wondered who it was he perceived was trying to hurt him in his torpid mind. The gun in his desk was not the only one he kept on hand. His briefcase was on a chair next to his desk. When he opened it, there was an identical gun in there as well. He said he carried it everywhere. In the car, he kept it on the front seat, and when his wife was with him, she sat in the backseat. I wondered why he was showing these guns to me.

The hospital's morale continued to deteriorate. Mr. Hall kept decreasing the number of direct caregiving employees while increasing the non-caregiving employees. He brought in a new personnel director, a new budget director, and a new business office supervisor. At the same time, he laid off nurses and technologists. Some Sisters liked him, because under his regime, the hospital made large sums of money, and he always presented himself as a savior of big catastrophes.

He had a lot of tricks in his bag. We had a test called SMA 12, during which a large autoanalyzer performed a panel of 12 tests per minute on a patient's blood. This cost us about five dollars per panel. We charged $12 to defray indirect costs. This was a laboratory test profile on a patient for a reasonable cost. One day, Mr. Hall showed up at my office with a calculator in his hand. He wanted to know how much it would cost if we charged these 12 tests individually. I added them up, and it came to $78.

He said Medicare would pay our usual and customary fees and suggested that we charge $78 for these tests, and he would enter a $66 credit for non-Medicare patients in the computer. Then he calculated that it would bring in $540,000 in extra revenue a year. He pointed out that because my compensation was related to lab revenues, there would be a lot of money in it for me as well. I explained to him that I could not agree to it because such an arrangement might be illegal. He said he understood, and that he was just trying to help the hospital and me, but this was not the end of it.

A few months later, one of the Sisters told me that Mr. Hall had told the leadership of the order at a Motherhouse meeting that I was a brilliant pathologist, but had no business sense and was extremely stubborn. Just one example of my hard-headedness supposedly cost the hospital $540,000 a year. The governing board of the Motherhouse was a group of elderly Sisters who had been cloistered for decades and did not understand modern hospital business. But, they did understand $540,000. When I heard that, I wondered whether or not he would have been the one to report me to Medicare if I had agreed to his Medicare scheme.

Slowly, hospital personnel became divided into two groups: those who toed the line with Mr. Hall and those of us whose main objective was giving good patient care. This also included the Sisters who worked in the hospital. For instance, the Director of Nursing Services, a young nun with an advanced degree, told me if they didn't get rid of Mr. Hall, she would leave the order.

The bright young Assistant Hospital Administrator, Mike Abell (Mr. Hessen's replacement), also refused to toe Mr. Hall's line. One day, Mike bought a new Cadillac. The next day, he found his car windows shattered. He had no doubt Mr. Hall's was somehow involved in this. One could feel his heavy hand everywhere. For instance, the Director of Personnel, Mr. Hall's flunky, tried to hinder me any way that he could. If a key person left the lab, he or she was replaced immediately in the old days. Not anymore. Now, the replacement was delayed weeks or months or denied altogether.

One summer, we went on vacation. Upon our return on a Sunday, I called my associate, Dr. Dickstein, to tell him I was back. He wanted to come over immediately to tell me about the crisis in the lab. Apparently, while I was away; Mr. Hall and one of his Assistant Administrators took the Laboratory Supervisor out to dinner. Afterwards, they took him to a nightclub girly show and got him pretty drunk. He was dancing with the partially dressed ladies of the house while they were feeding him grapes covered with sugar. They also told him that if he got some information that would help them get rid of me, he would be chosen to replace me as Lab Director and enjoy a very large increase in salary. What they did not tell him was that as a Laboratory Technologist, he could never be a Laboratory Director in an accredited hospital.

Their next move was to have that same Laboratory Supervisor try to fire anyone who was suspected of being loyal to me. He would come to my office and say we have to let so-and-so go. I would ask for the reason, and he would say—invariably—that he or she had a "bad attitude." Finally, I had enough of that, and I told him in no uncertain terms that I was the only one who could hire or fire. In addition, that termination recommendations had to be in writing with his signature with the specific reason for it noted. In this way, if a lab employee wanted to file legal action, they would know who recommended termination and why. This scared him off from any further attempts to get rid of good, hardworking people. Despite all of this going on, we provided accurate diagnosis and laboratory services. I tried not to take these problems home with me, but I still had many sleepless nights over the whole situation.

For a while things seemed quiet, but then Mr. Hall convinced the Motherhouse to order another audit of my referred lab work from our private lab to the hospital. We agreed that the auditors would have to be from a well-recognized accounting firm. They could examine our books in our office and the hospital's books at the hospital. Each party would be provided with a report of the results. The audit was done, but no results were ever forwarded to me. After our request for the results, we were told there was no report on the audit.

Later on, I found out that the audit had found that we had not only paid the hospital for every single test we referred to it, but that we had even overpaid. The auditors also pointed out that a public apology was necessary for what I had been put through and, in addition, it should be stated that I was completely innocent of any wrongdoing. The audit also stated that I should have recognition for my excellent job performance and should be asked to serve on the board. We found out that the results of the audit, which completely exonerated me, had been presented to the Motherhouse. The Sisters requested that the auditors not provide a written report of the audit results, but just a tape recording of the audit results presentation. We never had access to the tape and later were told it was "erased by mistake."

After all of this, Mr. Hall still maintained his position as number two in the hospital's administration, and he was also a "defacto" chief executive. None of his wrongdoings, which included a Medicare penalty for overcharging emergency room patients and an order to refund the money to the patients, were questioned by the Board or the Sisters.

I thought with the audit behind me, my troubles with Mr. Hall would be over. Not quite yet. We utilized auxiliary ladies for such clerical functions as filing or report dispatching. They were usually elderly, distinguished ladies who volunteered to help the hospital. I noticed that there was an auxiliary lady who was much younger and rougher looking than the average auxiliary member.

One day, our office manager followed me out of the lab and asked if she could meet me privately outside the lab. She wanted to tell me something, but not in my office. I didn't quite know how to take this request, so I suggested we meet in half an hour on a busy nearby street. When we met, she told me that the younger volunteer was not a volunteer, but a private detective hired by Mr. Hall to spy on me. She was trying to find out if I had stolen from the hospital, had a mistress, or participated in anything compromising. She was there for six weeks and couldn't find anything. The private detective told our office manager that if she helped her find some dirt on me, Mr. Hall would issue her double paychecks. I told her she should take the offer, because it was a lot of money and it wouldn't hurt my feelings. She said absolutely not, even if she got fired for it. She also said she got sick to her stomach when this undercover private detective made the offer and almost threw up in her face. She told the detective she would think about it.

When I told this to my attorney, he suggested I ask the office manager if she was willing to give an affidavit about this offer. She said she would, and on her day off, she went to my attorney's office and gave the affidavit. Afterward, my attorney said I had nothing to worry about.

He suggested we meet with the hospital's Chairman of the Board in his law office, not in the hospital. We gave the chairman a review, starting with the negative way the hospital responded to my recommending for Mr. Hessen (who had been unjustly fired) for a hospital administrator position. We discussed the FBI visit to my home, the audit of my private lab, and the offer by the undercover private detective to one of our key employees. He listened to us and said that I was a great asset to the hospital, and that he just couldn't believe what I was telling him. Then my attorney assured him we had legal proof for all this harassment. He thought it sounded like a nightmare from Communist Hungary. He said he would check into it and would get back with us. Nothing happened.

As some of this leaked out, a lot of doctors on the medical staff were very upset about my unjustified accusations, innuendos, and interference in performing our services. The chairman of the executive committee told me that the medical staff would not tolerate my continuous inquisition-type harassment, and that he would call a special staff meeting with only one agenda: "Makk vs. Hall and St. Anthony Hospital". He asked me to present everything, starting with Mr. Hessen, and to take all the time I needed. I knew the room would be bugged, but I decided to go ahead and present everything. In the meantime, I ran into the hospital malpractice defense attorney after church one day. He told me there would be some "skullduggery", but to hang in there. During the week of the special staff meeting, he and his senior partner came to the hospital and informed the administrator that if anything happened to me, they would no longer defend the hospital for malpractice.

As the special staff meeting approached, the pressure I felt was almost unbearable. I decided to give my presentation with all guns blazing. I almost felt like I had in 1956, when we had given our all to get rid of the Communist dictatorship in Hungary. When I finished my presentation, I got a standing ovation. Then a senior medical staff member, who was also the president of the American Academy of Family Physicians, made a motion to express no confidence in Mr. Hall. It was seconded and approved by a unanimous acclamation. Then the president of the medical staff went to inform the administrator. She asked what the ruling meant, and he told her that they would not bring any patients to the hospital as long as Mr. Hall was employed there.

This happened on a Thursday. Nothing happened Friday or Saturday, and Mr. Hall was still there on Sunday. Shortly after my family and I returned from mass that day, there was a knock on the kitchen door. It was my ally Assistant Administrator Mike Abell. He informed me that Mr. Hall and his Director of Personnel were no longer employed by the hospital. The Chairman of the Board gave each of them 30 seconds to resign or be fired. Mr. Hall wanted to be fired, and the personnel director resigned. When they emptied Mr. Hall's desk, there was an organizational chart showing him at the top as the CEO and everyone else underneath him, including the current administrator, Sister Francis Anne. Later, we heard that Mr. Hall went to another hospital, and within a year, he was fired at the medical staff's insistence.

The following Monday, there was great relief in the hospital. Employees stopped me and thanked me for spearheading the movement to rid St. Anthony's Hospital of this wrongdoer. The medical staff expressed their gratitude by electing me as their next president. I was also elected to be vice president of the State Medical Association. Later, I was also elected to the hospital board. I was honored with the hospital's Outstanding Service Award as the fourth recipient of this award in the hospital's a 100-year-old history. I also received the City of Louisville's Outstanding Citizen Award that year.

The hospital board members respected my opinion to the point that one of them publicly stated that he would ask me about medical matters and base his vote on my opinion. The Sisters from the Motherhouse were cordial, but never expressed any regrets for what I had been put through during the Hall regime.

Interestingly, Mr. Hall would reappear in my life. Nearly 10 years later, I got a call from him. He said that he had heard the hospital was looking for a chief financial officer and asked if I would give him a good recommendation. I told him I would give a truthful recommendation, as I always did. The phone clicked as he hung up, and that was finally the end of it.

Chapter XIX

A New Discovery

Finally, things truly settled down at St. Anthony's and the focus returned to providing excellent service and care. Indeed, much good work came out of our Department of Clinical & Anatomical Pathology. During the 1970s, we made a very exciting discovery involving liver damage and angiosarcoma in vinyl chloride workers. This discovery led to the development of a vital occupational health-screening program used in the rubber and plastic industries.

Frozen sections are important diagnostic tools used in pathology. They are examinations on tissues or parts of tissues immediately after surgical removal from a patient, frozen sections were important diagnostic tools used in pathology. The tissue specimens were quick frozen, and microscopic slides were prepared from them. The diagnosis was rendered within minutes and reported to the surgeon while he was still operating. One evening around nine o'clock, I got a call from the operating room that Dr. John Creech (a colleague at St. Anthony's) wanted me to do a frozen section.

Dr. Creech was a well-trained surgeon with excellent judgment. He was also the company doctor for the local BF Goodrich Company. This evening, he was doing an emergency operation on a Goodrich employee who had a bleeding stomach ulcer. He noticed an unusual lesion on the liver and took part of it out for frozen section diagnosis. After I prepared and studied the microscopic slide, I knew it was a cancer but not what kind. As I strained my mind and eyes, I realized that I had never seen this before in the liver. The more I studied it, the more I felt like I had déjà vu. Then I realized that it was similar to a cancer removed from the leg of a racehorse that was sent to me for a second opinion. It was a cancerous growth of blood vessels. That was it! A cancerous growth of blood vessels called angiosarcoma. I went back to the operating room and reported it to Dr. Creech. He very astutely asked how I knew it was angiosarcoma if I had never seen one before in the liver. I got my camera and took pictures of it through the surgical wound. I went back to my office and started to look for a liver angiosarcoma in my reference books. Finally, I found one case in a report on 50,000 autopsies.

The next morning, I did some more research on it, but information was very scarce. I did find out that there were 24 new cases a year in the United States or about one case in 10 million. Dr. Creech came by my office and wanted to know if the diagnosis was still the same. The answer was yes. I told him this was such a rare tumor that we ought to publish it as a case report. I also took photomicrographs, photos through the microscope, for future publication. We then became very busy and didn't get around to publishing the case report at that time.

The next December, Dr. Creech came by and asked if I remembered that case. He said that there was another case of angiosarcoma reported on a death certificate of a Goodrich worker. I suggested to Dr. Creech that he should have the company nurse go through the death certificates and pull out every case where liver cancer or angiosarcoma was listed. There were three more cases of people who had died in other hospitals. I contacted the pathologists at the other hospitals and asked them if I could review the microscopic slides from these patients' autopsies. The first pathologist I contacted asked if I wanted to review the slides because they were being sued, and I quickly said no. Then he suggested that they had sent slides out to three different places for second opinions, but they got three different diagnoses back. With these five cases, the incidence of angiosarcoma was 42 thousand times higher at the Goodrich plant than it was in the normal population. We were onto something now!

To me, this was very exciting, because when I had been a teenager, I had read a book about how a young English doctor named Dr. Arrosmith discovered that the water supply had been contaminated and fought with the authorities who wanted to suppress his findings. I thought that we might have had to do the same here, but we did not. Goodrich, on its own, immediately reported these findings to OSHA (Occupational Safety & Health Administration). Dr. Creech invited me to accompany him to the plant to meet with the top managers, and they also came to my office. They wanted to know how we could find out if there were more employees with angiosarcoma. I suggested a liver test medical screening program. Before we went into specifics, I asked them how much they could spend on such a program. The plant manager said he wanted to go first class and not worry about the cost. My recommended screening program detected two more cases. On worksite analysis, it was found that the only people who developed angiosarcoma of the liver were those who cleaned the chambers that produced polyvinylchloride from vinyl chloride monomers. Polyvinyl chloride was the basic ingredient of plastics, but vinyl chloride was a gas. The people who developed angiosarcoma were workers whose assignment was to get into the polymerization chambers and remove the polyvinyl residues that were attached to the chamber walls. In this process, vinyl chloride residues were released, and the workers inhaled them. The vinyl chloride molecule was similar to alcohol, and it had a similar effect on the workers as a drink or two.

Shortly after these discoveries, the New York Academy of Sciences had a symposium where Dr. Creech presented our findings. Suddenly, we were on the map. Dr. Hans Popper, the world-renowned liver expert, came to visit every second week. Dr. Lou Thomas, the chief pathologist at the National Cancer Institute, invited me to a conference with him periodically. I prepared a set of microscopic slides for each of these cases. A couple of months later, I picked up an issue of *The New England Journal of Medicine* and, lo and behold, there was an article on vinyl chloride-related angiosarcoma by Dr. Popper and Dr. Thomas. It didn't give credit to either Dr. Creech or me. It made it appear like these had been their own cases.

Another interesting episode happened when I took my cases to the Chairman of the Department of Pathology at the University of Louisville School of Medicine. We reviewed each case, and I asked him what he thought. He said if he were me, he would keep these cases in the bottom drawer of my desk and would not sign them out until I had retired, because there was so much liability that I would spend most of my time in courts if they came out. So much for that advice. I had already signed them out and never spent one minute in court. There was tremendous interest in our detection programs and our cases. I wrote an article about these cases and submitted it to the *Journal of the American Medical Association*, the most-read medical publication. The article was promptly published in October of 1974, and I received reprint requests from more than 40 countries.
Later, I described the pathologic findings of these tumors and submitted it to another prestigious medical journal called *Cancer*. My article entitled "Clinical and morphologic features of hepatic angiosarcoma in vinyl chloride workers" was published in January of 1976. Pictures I had taken in the operating room on our first case accompanied the article and were published in living color—a first for the journal.

I received cases from all over for consultation. I prepared a scientific presentation on the pathologic findings of vinyl chloride-related angiosarcoma of the liver and presented it to the James Ewing Society, which is now called the Society of Surgical Oncology, the most-prestigious cancer organization. My presentation was received with a standing ovation and many laudatory comments. I was then asked to join the James Ewing Society. It was a high honor, as I was one of only two members who were in the private practice of pathology and not an academic.

As time went by, other companies also wanted us to screen their workers for occupational diseases. In fact, at one time, we chartered city buses to shuttle workers to and from screening in our laboratory. This was a very lucrative enterprise for the hospital. Several doctors, including Dr. Creech, urged me to transfer the screening to my private laboratory, but I felt the hospital was a better place to conduct the screening. I also did not want a large private laboratory, because my training was best suited for hospital pathology.

Now that the occupational health screening program was well established, a symposium was held in Louisville with the objective of finding what we could learn from this crisis and what we could do to prevent anything else like this from happening again. From this meeting came a recommendation to create a large health-screening program offered to each worker in the rubber and plastic industry. I was asked to design and implement the program and act as the chief medical consultant for it. Eventually, it was funded by a one-cent contribution from each hour worked by the companies and another cent by the union.

I also designed a review process that consisted of studying the results of participating workers from each plant and comparing them to the company's results, which covered several plants, and then to compare these to the industry's results. This way, we could detect if there was a toxic health effect in a plant or company early, because some of the tests would show abnormalities, with higher incidence than present in other locations. We thought this would detect toxic effects before clinical illness occurred so that we could hopefully reverse the damage.

I did this for years, and it became the largest blue-collar worker health screening program to date. We hardly detected any toxic effects. The most important results were that, through participation in the screening program and health education, the workers enjoyed better health than the general population.

Chapter XX

New Horizons and Declines

The constant training of our staff bore fruit. With our group of motivated and smart employees, we were able to set up complicated laboratory tests which were not available in other local or regional hospitals. Pretty soon, such tests were sent to us from other areas and even out-of-state hospitals. This not only added to the prestige of our department but also produced good income to the hospital.

Unfortunately, Dr. Ed Fadell, my colleague I had met during my residency in Texas and friend who first put me in touch with St. Anthony's, suddenly passed away. At the time of his death he was head of the pathology department and Director of Laboratories at Methodist Hospital in Louisville, I was offered his position shortly thereafter. Methodist Hospital was a more advanced hospital and had a better medical staff than ours. The offer was very attractive, but I had built our pathology department and laboratory from scratch. When I came to St. Anthony's, all there was at the time was a concrete area where the lab was going to be. I did the planning and helped build it from brick and mortar into a well-organized and successful department. We trained most of the staff, too. This was my baby, and it was like family. I felt I owed my loyalty to St. Anthony's, even though Methodist was a better offer.

I came up with the idea of uniting the pathologists from both hospitals and becoming the head of the pathology department and laboratories in both for a year. This way the expertise of both pathology groups would be available in either hospital, and later on, we could bundle tests and perform them in one or the other lab, resulting in significant savings. I proposed a one-year trial period, and both hospitals agreed.

The first week at Methodist, the administrator asked me to have lunch with him. He was very much an in-charge person. At St. Anthony's Hospital, we had an active industrial workers health screening program, which was quite profitable. He wanted me to transfer it to Methodist Hospital. I told him no, and I could see from his reaction that he was not used to people saying no to him. From then on, he was very distant.

During lunch, he mentioned that I should let one of the pathologists go, because her diagnostic work had been questioned and because the doctors did not trust it. I thought he was reaching too far. Furthermore, the pathologist in question was good and had been well trained. He also mentioned that he would be very much in favor of me hiring Dr. Meadors, who had applied for my position when I had been selected. This was a big surprise, because I was an independent contractor and it was up to me to choose my associates. They worked for me or with me, and their compensation came from me, not the hospital. He had way overstepped his authority. I told him I would check on the work of the pathologist in question by coming in late evenings and early mornings. If I found that her work was faulty, I would terminate her. After I checked her work, I found her to be a smart, motivated pathologist, and I doubled her compensation.

The administrator still would not let up on hiring Dr. Meadors. He said he had run into him in an airport, and he was still very much interested in working at the hospital and would have been happy to take a position under me. Next, the administrator offered to pay for his expenses for an interview. We needed a pathologist, so I agreed. He was okay but not too impressive. I offered him the job anyway. After we started to work together, I realized he was not a very good pathologist at all. He was very uncertain and very religious. Most pathologists' desks were full of reports and microscopic slides. He had only a Bible on his desk.

One morning, I came in to check out the other pathologist's work around 6:00 a.m., and I noticed the light was on in Dr. Meadors's office. This was unusual, because he usually came to work around 8:00 a.m. I thought he had left the light on, so I went to turn it out. His door was locked, and when I knocked on the door, he opened it. I saw that he and the administrator had been meeting behind locked doors. Both seemed embarrassed. The administrator kept assuring me that it was not a business meeting, just a social call, because he had come to the lab for blood work and had seen Dr. Meadors. After I checked this, I found out that the administrator had had no blood work that morning.

The end of the one-year trial period was approaching, and it was time to rework my one-year contract to a continuous one. To every verbally agreed upon draft, the administrator later had some Mickey Mouse objections. Then he named the Chief Financial Officer and an Assistant Administrator to negotiate with me. They told me that the medical staff wanted me and to tell them what they needed to do to keep me there, but the administrator still had some objections. Finally, I'd had enough of these objections. I also felt that I belonged at St. Anthony's. I had built the place, and it was like a professional home to me, the employees like family, so I wrote to the Methodist Hospital administrator that I was no longer interested in pursuing negotiation and would be leaving. The medical staff was very upset. Like me, they felt that the administrator really did not intend to enter into a contract with me. He wanted Dr. Meadors as chief pathologist and laboratory director. This could be done if I was no longer there. I paid a price again for standing up for what was right.

Dr. Meadors was a very submissive, insecure person—perfect for an administrator with an aggressive, domineering personality. Shortly after Dr. Meadors got the job, the pathologist whom the administrator wanted me to fire was terminated. As time went by, Dr. Meadors's inability to cope became obvious, and he was let go.

During these times, I received many public honors. I was elected to serve on the board of the regional Red Cross, Junior Achievement and St. Anthony Hospital. I also served on the new product evaluation and cancer community of the College of American Pathologists, and became a member of the Council of Laboratory Administration.

St. Anthony's was doing fine for a while, but as the hospital industry became more competitive and most hospitals acquired lay professional administrators, it became obvious that St. Anthony's just couldn't compete. For instance, our hospital was the first to have a cardiac intensive care unit, but we never developed a heart surgery program even though we had the surgeons and the will.

Chapter XXI

Health Crisis

In 1975, we purchased a 305-acre farm outside of Louisville in La Grange. We named it *Six Acorn Farm* because "Makk" in Hungarian means "acorn", and there were originally six of us. Our whole family has enjoyed it over the years even with the hard work involved. We have a small thoroughbred broodmare operation, raise cattle and grow hay, The children have helped over the years (and now the grandchildren) with maintaining a vegetable garden, fruit orchard, vineyard, as well as bee hives and honey production. Carolyn enjoyed her Arabian horses and preparing homemade fruit preserves that won many blue ribbons at the State Fair. I have loved the physical labor involved and particularly working with the hay.

In 1988, we were going to have an Easter eve dinner at the farm. One of our mares went into labor that afternoon. After we delivered the foal and got ready to give post-foaling care, I developed a bad shaking chill and became so weak that I could barely move. Carolyn put winter coats and blankets on me and turned the thermostat as high as it would go, but I was still freezing and shaking. She wanted to call an ambulance, but I resisted it. At around one o'clock in the morning, we decided to go back to Louisville to be close to my doctor and the hospital. I insisted on driving and wouldn't move from the driver's seat. Carolyn was scared to death, but we got home. By the next day, I felt even worse and did not feel I could go to Easter Sunday mass.

Andrew, our youngest son, was not sure I was that sick. He thought maybe I just didn't want to go to a long mass. He put a slice of ham on the kitchen counter. He told Carolyn that if the ham was still there when they came home, I really was sick, but if it was gone, then I was well enough to sneak downstairs to eat the ham. When they came home, the ham was still there, so according to the "Andrew ham test," I was really sick.

Laz, our eldest son, was a medical student, and because I was getting worse, he would come to my room periodically. He told me that he needed to practice taking blood pressures and asked if I minded that he practiced on me. Of course I didn't mind. I did not know he was constructing a medical record with the blood pressure and pulse measurements. By the late afternoon, he called my doctor and presented my case. My doctor came over right away, took one look at me, and drove me to the hospital.

By the time we got to the hospital, I was in shock, and I did not remember anything later. When I eventually woke up, I was in a hospital bed with IVs everywhere. For a couple days, despite the best efforts of the best doctors, my diagnosis remained a question. I got so weak that it took me fifty minutes to turn from one side of my bed to the other.

It turned out that I had septicemia. A germ called staphylococcus aureús was growing in my blood vessels and on my heart valves. This was extremely difficult to treat, and shockingly, the survival rate was only 50 percent. After I was released from the hospital, I was on intravenous antibiotic therapy at home for a while. I never had any doubt that I would recover and get back to work and play tennis again. Thank God and my excellent doctors, because I recovered faster than anyone had anticipated. However, the recovery was not complete. One of my heart valves had been severely damaged and had produced circulatory problems, but I could still function well enough and became active again.

Several years later in 1991, I developed chronic fatigue. I went to see my doctor, and he ran the same tests they usually run. Ever since my hepatitis, some of my liver function test results are abnormal, and they were this time as well. My doctor jokingly told me that I had always been hyper, and that now I was getting to be normal. Actually, my liver function tests showed that my liver was getting worse which explained my persistent fatigue. I had a liver biopsy done to be safe. When I looked at my slides the next day, I saw the worst diagnosis: chronic active hepatitis. It was worse than a death sentence, because all patients died from chronic active hepatitis at that time.

I asked one of my partners to review my slides. He came in and put them on my desk and could not speak. I asked if he agreed with the diagnosis, and he just bowed his head in agreement. Now I had to tell Carolyn and the kids. After the initial shock, I decided to search for ways to stay alive. There was a new treatment with interferon. It was still somewhat experimental with only a 25 percent success rate, and it was not yet known if this 25 percent experience temporary or permanent success. Regardless, it was my only option at that time, and my liver doctor put me on the medication. I had to give myself the interferon injection twice a week and check my blood count once a week. The medicine had a tendency to kill the white blood cells, which would have made me very prone to infections.

With my reduced immunity, I kept on working, but I avoided conferences and went to the cafeteria when they were the least busy to avoid catching a cold or infection. I usually checked my blood count on Saturday on the way to evening mass, because the lower attendance gave me a lower chance of getting an infection. After church, I picked up my blood count report. If the white blood count was okay, I usually took Carolyn out to dinner no matter how badly I felt. If it was low, we just went home and tried to cheer each other up. Our faith in God was tremendously helpful, and we never despaired.

After a while, it became obvious that this treatment not only did not help but made my condition worse. I got real sick and suffered constant pain, fatigue, and itching. One day, I was so weak that I could not get up. We tried to reach my doctor urgently, but to no avail. He wasn't returning my desperate calls. He was the doctor with whom I had interviewed for a position at the university that he ended up taking. He actually had his first meal in Louisville at our home. Just before his office closed for the day, Carolyn reached his secretary who was also his wife. After Carolyn's respectful pleas, she said the doctor was busy, and that he would return the call when he could. Finally, he called around 9:30 p.m. and advised me not to go to crowded places like supermarkets, where I might catch a germ. When I told him I was so weak that I could not even get out of bed, he didn't even suggest that I come into the office to be checked over.

That night, I realized I didn't have that much longer to live. The next morning—this was June—I called the Mayo Clinic's Hepatology Division. I found out that the next available appointment there was in November. I told the scheduler I wouldn't live that long and asked her to get in touch with me if they had a cancellation. Fifteen minutes later, the phone rang. It was the Mayo Clinic notifying me that they had had a cancellation for Monday morning at 8:00 a.m. I immediately told them that I would be there.

Sure enough, I was there in Dr. Rakela's office at eight o'clock on Monday morning. He was one of the world's most prominent liver disease experts (hepatologists). He very kindly mentioned that he was familiar with my pioneering work in vinyl chloride-related liver tumors. After a careful history and exam, he advised me not to take interferon again, because it might kill me if I took more. My hepatitis was the kind that not only did not respond to interferon treatment but was worsened by it. So much for the hepatologist in Louisville whom I had helped to get hired at the university. Later on, I found out he was doing clinical trials for the drug company that was making interferon and got paid a nice sum of money for each patient. I never figured out whether it was ignorance or greed on his part that led me down such a dangerous path. Dr. Rakela recommended other medical treatment modalities but also politely explained that there was no guarantee that any would work.

As time passed, I got a little better. We could not establish whether my improvement was the result of not taking Interferon or taking the new medications. I was not well, but I was stable for the next three to four months. Then around Christmas, my fatigue got worse. In January 1992, I signed up for a weeklong workshop given by the Mayo Clinic on new laboratory procedures in Boca Raton, Florida. It was so nice to be in the warm sunshine. Some of the Mayo faculty played tennis in the evenings and invited me to join them. The first night, we had a great game. That night, we had a seafood buffet. I really enjoyed the fresh seafood. The next day, I had trouble following the lectures. My movements became uncoordinated and sporadic, and when we went to our tennis game, I couldn't hit the ball at all. Finally, I excused myself from the game. On another night, the hotel had a two-pound lobster special, and I ate one. The next morning, I could barely shower. I also had trouble walking and could not concentrate on any of the lectures.

After the meeting, we arranged to stay for a week in an apartment at the Everglades Club in Palm Beach with help from Mrs. Bender. I had trouble driving to Palm Beach, walking, negotiating the steps, and carrying on a coherent conversation. Our son, Andrew, came to visit with some of his Tulane polo team members. I am sure they thought that I was strange. One of our other sons, Steve who was in orthopedic surgery training, also came to visit us. He thought I was getting real sick, and we had better get back to Louisville. By this time, I was losing consciousness at times. What we did not quite realize was that all of the seafood had been bad for my liver, and that I had developed a condition called hepatic encephalopathy (liver failure affecting the brain function). I was not capable of driving anymore, so Steve drove us.

On the way home, I got even worse, so Steve and Carolyn decided to take me to Vanderbilt Hospital in Nashville. In Nashville, I became a little more alert and begged them not to take me to Vanderbilt and to try to take me to Louisville where all of my doctors and medical records were. We did make it home eventually. My primary doctor lived next door, and I saw his car in the driveway. I asked Carolyn to call him and ask him to come over to check on me. After he examined me, he told Carolyn that I was so far gone that no one in Louisville could save me, and that the Mayo Clinic might not even be able to if I didn't get there by morning.

There was no commercial airline flight overnight to the Mayo Clinic. We started to call air charter companies, but in each case, we got a recording saying to call back on Monday or leave a message. Then I realized that one of the attorneys in our tennis group had chartered a plane to fly to his cottage in Cape Cod, but we couldn't get a hold of him either. Then we called Bill Stogdhill, another tennis friend, to see if he would know how to get in touch with our mutual tennis friend. He said he would see what he could do. About 15 minutes later, another tennis friend, David Jones (the CEO of Humana) called and told Carolyn: "Betty (his wife) and I love Laszlo, just as you do. My jet is getting ready to land at the airport and will be ready to take you and Laszlo to the Mayo Clinic in Rochester, Minnesota right away. Bill Stogdhill will pick you up and take you to the hangar."

We were airborne right away. We landed in a snowstorm and almost hit a fox on the runway. The pilot had kindly ordered a taxi for us. A winter snowstorm was coming, and there was a fuel truck waiting to refuel the plane so that they could get out before the airport was closed. I was so confused that I thought the big fuel truck was the taxi. As I tried to get into the cab of the truck, the people on the ground pulled me back and took me to the taxi. The ride from the airport was slow on the icy roads. By this time, it was nearly 3:00 a.m., and we decided that we would check into the Kahler Hotel and go to the clinic when it opened at 7:00 a.m. rather than go directly to the emergency room. I was seen immediately. The doctor took a look at me and said he was not going to take up my time by examining me. I needed be in intensive care and was sent over there right away. He called for a patient transport and wrote my admission orders. In no time at all, I was in the ICU receiving treatment. In addition to my hazy memories, I put together these events from what Carolyn and others have told me. After about four days, my mind started to clear up, and I researched hepatic encephalopathy. There were four grades of it, and mine was around grade two and a half. This was good news, because people who advanced to grade three were not eligible for a liver transplant.

One of the first things I did after my mind cleared was call David Jones to thank him for my lifesaving flight. I asked him to let me reimburse him for the flight, but he wouldn't hear of it. Then he asked me to do him a favor and call him when I was released so that he could send the plane for me. Of course, I did not call him since we could fly commercially.

As I was recovering at home in Louisville, my Mayo Clinic physicians started to talk to me about a liver transplant. I had a couple of strikes against me. I was 62 years old, and at that time, the Mayo Clinic had only transplanted a liver into one other person who was older than 60 years. Another strike against me was my damaged heart. Nevertheless, they felt that I should be worked up for a transplant to see if there was anything that would preclude me from having one. Nearly every medical specialist and even a psychiatrist examined me. After my interview with the psychiatrist, he asked me if I had a letter to my family in case things didn't turn out the way we all had hoped. I told him that if I got the transplant, I planned to walk out of there, and for this reason, I had no such letter. He cautioned me that it was a big stress and that the recovery was long, boring, and depressing. I told him it was okay, because I would have time to figure out which stallions to mate with my mares. He cracked up and said, "Heck, you'll get through this without bending a hair."

I learned that after all these consultations were completed, the transplant candidate's records were presented to a multi-specialty medical committee to decide who was accepted to be transplanted and then put on the waiting list. I found out the time and location of the conference meeting, and I happened to be at the Mayo Clinic at that time. I sat on a bench near the conference room and waited. I was hoping there would be a doctor I knew who could tell me if I had made the cut and made the list.

Finally, the conference room door opened, and the doctors began to leave. Luckily, one of them was my doctor. He told me that I made it and that I was being put on the transplant waiting list. He also told me I was lucky for having a B+ blood type, because there were only five or six people ahead of me. If I had had an A+ blood type, there would have been 62 people ahead of me. I was grateful to God that I was not an A+, because I would not have lasted through a long waiting period.

I flew home happy with the new hope that I might live. I had to go on a specific no-protein diet. We had to order the food from Rochester, New York, and it tasted like cardboard. We had to go back to the Mayo Clinic for frequent checkups. During one of these visits, Carolyn and I had to attend classes to learn what to expect. On one of the visits, I was sitting next to a farmer whose wife was also on the waiting list. I was reading an article in *The Wall Street Journal* about a pig liver being transplanted into a human. The farmer told me I seemed like a nice fellow, and that if I decided to have a pig liver transplant, he had two thousand pigs on the farm in Iowa and that I could come and pick one out. At the end of the article, it said that the patient promptly died. In the fall, I also had to take a horrible-tasting antibiotic cocktail to sterilize my intestines. My fatigue was so terrible that I was totally exhausted by the time I finished my morning shower.

Before the transplant plans were being made, we started to build a new house on the farm where we had planned to retire. The question came up about whether we should stop or keep building the house. I decided we would keep going, believing that I would end up healthy. Carolyn did a magnificent job in planning and decorating. Our architect was the renowned Stratton Hammon, and today, our house is featured in a book about his work.

I wanted to give Carolyn a nice housewarming present for her tireless efforts with me and the house. Once, at the Kentucky Derby Museum, we both admired a horse painting of the three Arabian stallions from which all thoroughbred horses originated. The curator told us it was by a young, talented horse painter who lived in Lexington, Kentucky. I got her name and address from the curator, because I wanted to commission her to do a painting of our Arabian stallion, Snoopy, to give to Carolyn as a housewarming present. Snoopy was sort of a member of our family. Our children helped with his birth as well. He was a gentle, smart, very friendly, and beautiful horse. His conformation was perfect, and he earned the legion of honor designation in the show ring.

When I contacted the talented painter Julie Wear about painting Snoopy, she told me she knew him from the horse shows. She was interested in painting him, but it would be nine months before she could get to it. She expressed interest in observing Snoopy in his natural environment. On her next trip to Louisville, she said she would stop by the farm. That was on the Fourth of July. That day, we planned a picnic and invited a lot of officers who had just returned from Desert Storm and had been stationed at Fort Knox. We wanted them to raise the flag on our new flagpole before anyone else. This made the farm extremely crowded.

On that visit, she saw Snoopy happily running around the pasture and took dozens of pictures of him. She loved one of our Jack Russell terriers, Happy Jack, and took a lot of pictures of him, too. When I asked her when we could expect the painting, she said nine to 10 months. I told her I was on the waiting list for a liver transplant and may not live that long. I asked her if she could speed it up. She promised to try. About four months later, she called to say the painting was ready.

She and her husband were coming to Louisville and offered to bring the painting. We agreed to meet at the farm and hang the painting in the not-quite-finished house. The painting was spectacular. In addition to Snoopy, she included Happy Jack in the painting, and she used the pasture we viewed from the living room window for the background. I could not wait for Carolyn to see the painting.

I went to our home in town, and told her that we needed to eat at the farm that night. She didn't think it was a good idea to have our dinner in a house that was not quite finished, but I prevailed. She went through the living room four or five times without noticing the painting. Finally, I could not take it any longer and asked her to look above the fireplace. She was ecstatic. I knew right then—whether I made it through the liver transplant or not—the painting would always bring a smile to her face. It was amazing how love superseded illness. As I look back now, I do not know how I could have done so much when I had advanced liver disease. I practiced medicine full-time and was active in my many civic organizations while overseeing the construction of a new house with all of its frustrations.

In November of 1992, I started to go downhill despite all the medications. My fatigue became so extreme that I could not keep my eyes open, and I had great trouble concentrating; however, I kept on working. I was sure that my associates were kind enough to look over my shoulder to make sure I hadn't made any major mistakes. About this time, the Mayo Clinic called informing me that I had moved up to second or third on the waiting list. They suggested we move to Rochester to an apartment near the clinic in case I got a donor. That way, I would be available to receive my transplant, and bad weather would not keep me from getting there from Kentucky. We made plans to move up there after Thanksgiving. In the meantime, our house neared completion, and the builders wanted to be paid. I would have preferred to wait until the house had been completely finished, but as I was deteriorating, I was afraid I would get so incoherent that I wouldn't be capable of settling the bill.

After the Mayo Clinic call, I went by an air charter company and explained my situation to the owner—that I may need emergency transportation for my transplant. He was wonderful and provided me with his manager and his chief pilot's cell and home phone numbers. He informed them that they might expect a call from me, and when it came, they needed to get airborne pronto. Andrew made up a phone call list just in case I would go for the transplant. Consequently, if the time came, each family member would call another family member and so on. We had our belongings packed in suitcases under the bed so that we didn't have to pack if the call came.

After Thanksgiving, we were getting ready to pack and move to Minnesota. Before we knew it, December was upon us.

Even though my energy level was low, Carolyn and I attended the Louisville Bach Society's annual performance of Handel's Messiah as we always did. One our favorites, the Messiah spiritually charged us both and was made more special this year because my daughter-in-law Carolyn sang in the chorus. It was the first weekend in December.

By the time that Monday came along, my fatigue was so bad that I did not have enough strength to walk to the parking lot from my office. On my way out of the department, I stopped to talk to the employees so that I could rest a bit. As I was leaving the front desk, a carrier brought some microscopic slides from another pathologist for a second opinion. I told my secretary to leave them on my desk and I would study them the next day. Then I thought, *Somebody somewhere is waiting to find out if the slides show evidence of cancer or not.* I worked my way back to my office to study the slides and called the referring pathologist to tell him that it was not cancer. Then I started out for home again. When I got home, I just dragged myself up to the bed to lie down. Caring and loving, Carolyn said, "You will get the call tonight that they located a donor." She had said this before, but this time, she sounded very convincing.

December 8th is a major feast day for Catholics—the Feast of the Immaculate Conception—a feast that held great importance to me and Carolyn. December 8th was also a very special day to me personally, because I had the good fortune of arriving in the United States on that day. We used to celebrate it with a party, but there was no party this year, only a fight for survival. An attorney friend of mine scheduled a mass to be offered for my successful liver transplant on the Feast of the Immaculate Conception. A neighbor offered ahead of time to take us to the mass, but we were not able to make it. This would all prove to be a blessed coincidence.

The phone rang at 3:30 in the morning on Tuesday, December 8, 1992. It was the Mayo Clinic notifying me that I had a donor, and that if I could get to the Mayo by 10:00 a.m., I could get the transplant. I called the owner of a local charter airline and was told we could be airborne in 40 minutes. Our youngest son Andrew was home and initiated the phone tree with the brothers. I jumped in the shower. As I was showering, I started to wonder if the call had been real or if I had dreamed it. I jumped out of the shower and called the Mayo Clinic to see if the call had indeed been real. They assured me that it had been real. Andrew couldn't find his suitcase, so he just grabbed an armful of clothes from his closet, and we jumped in our car and sped to the airport. It was a little surreal, because there were more than 10 air charter people in Santa Claus hats. Then our children and daughters-in-law arrived driving 80 miles per hour when the speed limit was 30.

As I looked at my loving family, I suddenly felt like a commander going into the battle of his life, and my family's love would help me to win. Only six of us could fit in the jet. Chris and his wife Carolyn remained behind and jumped into our Jeep Cherokee and started the 725-mile drive to the Mayo Clinic. We were airborne just as promised and made the trip in less than one and a half hours. At about 30 minutes outside of Rochester, the pilot came back to the cabin and asked me to give him three other possible landing sites, because a winter storm was moving in and the Rochester airport might be closed. I told him Rochester, Rochester, and Rochester; but if the airport was closed, he could land anywhere near there. Thankfully, we made it to Rochester. The pilot had taxis waiting for us, and we were on our way to the Mayo on icy roads.

We didn't know exactly where to go in the hospital so we sent Andrew in to find out. He found a patient area and spoke to a staff person. When he told them he was there with his father for a liver transplant, they thought he was the transplant candidate and tried to put him on a stretcher. He finally got it across to them that he was just the son and not the transplantee. They came to get me, took me up to the transplant floor, and immediately started working on me. I had all kinds of tubes and IVs inserted into me, and then I was given enemas—one after the other—nearly until transplant time to sterilize my intestines. As 10:00 a.m. a nurse came in and informed us that the transplant had been postponed, for the airport had been closed because of a winter storm in the area from where the donor organ had originated. I knew that there was a time limit after which the donor liver would no longer be viable for transplant. So we had one more hurdle—the weather. After all these preparations, we realized that the transplant may or may not happen. The nurses kept on giving me the enemas and making preparations for surgery just in case the liver got there while it was still transplantable.

About four hours later, the good news came that my liver was on the way, and my transplantation was expected to start around four o'clock that afternoon. A solemn happiness descended upon us. Finally, after six months of waiting, we had overcome all of the hurdles, and it was really happening.

They were still giving me the enemas incessantly. I received holy communion from a female Episcopalian minister because the Catholic priest was not available. As the final minutes of waiting were clicking down, Chris and his wife suddenly appeared at the door. They had driven 725 miles in 11 hours. I had no idea how they had managed the trip, but it seemed like a miracle to me.

My transplant doctor, Dr. Ruud Krom, was supposed to be the best in the world. I thanked the Almighty for that, too. The transplant orderlies came, and as they were transferring me to the gurney, the television showed U.S. troops moving into Mogadishu. As we approached the elevator, Chris yelled out, "Dad, break a leg!" All of us started to laugh.

As I was being wheeled down the hall, I replied, "Keep smiling!" while I gestured with two thumbs up.

The operating room was extremely cold, and everything was shiny stainless steel. That was all that I remembered as they put me under. That said, the story about the operation originates from information received from my family and the transplant team. After 12 hours of operating, I still had a bleeder around the liver that could not be located. I was nearly dead and could not take any more surgery or anesthesia. I needed to get off the operating table in such a hurry that my abdomen was not even closed, just bandaged. Exhausted, Carolyn and the family, who had been up for 36 hours by now, were told to go to the hotel. The operation had to be stopped before it could be finished. They were going to try to stabilize me to the point where my operation could be completed. They did not say, and they did not need to say what would happen if I could not be stabilized.

By now, I had received 16 transfusions of blood and blood products and lots of IV solutions and medications. There's no real way to express what my family went through during this ordeal, particularly because my doctor sons knew well what it meant when a patient was removed from the operating table before the operation had been finished.

Fortunately, they were able to stabilize me so that I could be taken back to the operating room to attempt to complete the transplantation. Two hours after my family had left for the hotel, they were phoned to come back to the hospital, because I was being taken back to the operating room. They rushed over to the hospital, and after three hours of surgery, my transplant was completed. My family was then told that I would be kept under sedation for two to three days. Before they returned to the hotel for some rest, they were allowed to see me. I was so swollen that they could hardly recognize me.

I was warned before surgery that a lot of people have nightmares and hallucinations after a liver transplant—often about funerals. I definitely had my share afterward. Most of mine were about a funeral procession of tiny people being led by a dog with a crown on his head, marching around our vegetable garden. Either way, my family wondered if I would ever wake up and if I would recognize them when I did. Those days of waiting were like an eternity for my beloved family.

Finally, after three days, a nurse came to the waiting room and told my family that someone wanted to talk to them. As they came into my room, I suddenly blurted out, "What's going on in Mogadishu?" It was the nicest thing they could have heard. They knew I had not lost my brain function, because I had recalled the last thing on television before I was taken to the operating room. Eventually, I recognized everybody, and it turned out that I had no other side effects besides a foot drop (an inability or difficulty in moving the ankle and toes upward). I did also lose my sense of smell, but that is not always a bad thing for pathologists.

Postsurgical transplant studies showed that my new liver was doing well. Eventually, all the tubes going in and out of my body were pulled out. The last to be removed was a large tube that had been inserted into my jugular vein. That day, I underwent tests all day and could not eat anything. I had just started to eat dinner when one of my doctors came to pull the catheter out of my neck. For this procedure, one had to lie flat on his back, and as the tube was pulled out, pressure had to be applied with sterile gauze to prevent air from being sucked in. I was really hungry and wanted to eat, so I asked him not to make me lie down. I told him we didn't need a nurse, because I could hold the gauze to my neck. He relented and pulled the tube. I did not realize that I was experiencing a bad tremor in my hands. I then heard a noise like rainwater running down a down spout and suddenly felt excruciating chest pain down both arms and fingers. Before I lost consciousness, I thought that I was having a massive heart attack. Poor Carolyn was sitting by my bed. She thought I was gone right there and then. Suddenly, all hell broke loose. My doctors and nurses ran in, put a resuscitation board under me, declined my head lower than my body, and immediately took me to intensive care. They took off so fast that they yanked the phone, which was in my bed, off the wall. Nobody had any time to explain to Carolyn what was going on.

After a while, I started to hear banging noises that became louder and louder. Then I heard orders given in rapid succession by the director of transplant services and the head nurse. When I came to, everybody seemed relieved and delighted. Then they explained to Carolyn that I had had an air embolus that was caused by sucking air into my jugular vein when I did not lie down flat. Because of my hand tremor, air had gotten in my heart and had blocked the circulation in my heart, but luckily, the air did not get into the blood vessels and pass to my brain, because that was usually fatal. I felt so sorry for my darling Carolyn that she had to go through all of that.

The main reason I survived was because of God's help, and all my doctors and nurses had arrived there instantly, even though it was seven in the evening when it had happened. I did not realize at the time but learned later that 83 percent of air embolisms were fatal. I was put on oxygen after that; however, regardless of how much it was increased, my fingers and lips remained blue, and my blood tests showed that my oxygen saturation levels were getting low. My doctors were considering a tracheotomy to help remedy the situation. One day, my son Laz was sitting near my bed when he heard the oxygen whistling and went to the wall where the oxygen outlet was only to discover that the air was being pumped into the air instead of my lungs. He unscrewed the connector and found a crack in the oxygen outlet. The maintenance man replaced it shortly thereafter, and it saved me from having a tracheotomy.

While I was recovering, a new doctor came to my room, introduced himself, and started to examine my legs. I did not know who he was, because at this point, there were so many doctors coming in and out. Then he started to question me about where the bulldozer had crushed my legs. I tried to assure him that I had never sat on or even had been near a bulldozer at anytime in my life. He was pretty insistent, and I thought my nightmares had returned. Finally, he asked me my name again and said he was sorry. He was in the wrong room. Another doctor also came to see me. He was the CEO of the Mayo Clinic. He said his friend, David Jones, had called him to make sure I had everything I needed. He gave me his card and told me to call his private number if there was anything I needed. I thanked him and told him about the outstanding care I had received from the doctors and nurses.

The air embolus set my recovery back some, but when Christmas approached, I moved out of the hospital to a hotel that was connected to the hospital by an underground tunnel. I was still confined to a wheelchair, and Carolyn had to push me everywhere. Often, I had to go for tests, and on the way back to the hotel, she had to push me uphill on a carpeted area. Even now I don't know where she got the strength to do that.

Finally, I was able to walk on my own. Patients after liver transplants have diabetes, and I was no exception. One day, I was walking into the restroom when I lost consciousness and walked into a wall. I smashed my face and passed out. It turned out that I had fallen into a diabetic coma. After that episode, I started to feel well. My energy returned, and it was time to fly home. We had a long list of instructions—no visitors and no lifting even the slightest amount of weight. Visitors had to stay away, because I was so immunosuppressed that the slightest exposure to germs could have had catastrophic consequences. We were also instructed to have the air ducts in our house cleaned and disinfected.

Flying home was not as easy as being wheeled around in the hospital. Not long after that, however, we decided to move to our new farmhouse. It was isolated and very clean because it had never been lived in. I was cautioned not to go near anything that could contain mold of any kind. Most molds were innocuous to normal people, but if I was exposed to them, it could become fatal for me because of my immunosuppression. Carolyn had a very sensitive nose, and one day, she was sitting near an air duct and said something smelled like spoiled meat. After she opened the duct, she found pieces of bread and bologna covered with green mold. Here we were thinking that we were in a clean, safe house, and we had air ducts blowing mold all over the house.

While I was still recovering, some of our mares were getting ready to foal. Most would foal okay, but if there were any complications and the foal was not removed immediately, it could often be fatal. The mares had to be checked around the clock to see if their labor started and to assist them if necessary. Our farm manager was actually exhausted from doing this. I volunteered to check on the mares in the evening so that he could catch up on some rest. Our horse barn was on top of a small hill, and because I was not supposed to drive, I decided to walk up there. I was still weak, and my balance was off. The deep snow didn't help either. I decided to hang onto the fence as I tried to walk. At first, I could walk only as far as one or two fence posts, then three and four, before I had to stop to rest. I put a mask on my face before I entered the barn. This was great rehabilitation. When I went back to the Mayo Clinic for my three-month checkup, they couldn't believe what good condition I was in.

Before that checkup, I had an emergency return to the clinic. After I returned home from the transplant, I had weekly blood tests to check on my liver status. I started to feel tired again. All my tests came back showing more diminished liver function than the week before. When I called the Mayo Clinic, they felt that I was having an acute rejection of my new liver and advised me to fly up there immediately. After extensive studies, they figured out that I was not having a rejection but a toxic reaction to some of the many medications I was on.

When I was evaluated for the liver transplant, one of the tests I underwent was called bronchoscopy and bronchial washing, which involved putting a tube in my windpipe to see if there were any lesions and to put some saline into lavage the windpipe. One of the tests they did on the saline cultured it for bacteria and fungi. My bronchial washings grew one colony of a fungus called Aspergillus, which was present in the soil everywhere and only occasionally produced illness. When it did, however, the culture was loaded with them.

This one colony wouldn't have been of any concern clinically in people, because their immune system would usually take care of it; but my immunity had been severely suppressed so that I didn't reject my new liver. Therefore, it was decided that I receive large doses of antifungal medication, which not only controlled the Aspergillus, but also started to kill my transplanted liver. After the culprit drug was discontinued, I recovered. I had strict instructions again to stay away from moldy places like the barn.

After we got home to the farm, there was an urgent need to take care of a horse problem, but we could not get a hold of the vet. The problem involved our Arabian stallion Snoopy. He did very well in the showing in western pleasure classes. The national championships were coming up, and the best trainer accepted him for training and prepared him for the show. As a result, we had to send Snoopy to Texas. We had only hours to get him ready because a horse semi-trailer was coming through our area on its way to Texas.

We had been planning to breed Snoopy to one of our mares. At this time, we only had one large thoroughbred mare in heat, and Snoopy was too small to reach her. First, we had to open the vagina by giving local anesthesia, cutting it open, and putting a large stitch into it—a process called the Caslick procedure. With no vet available, I decided to do it myself. I put on triple masks to hopefully prevent me from inhaling any germs. We tried to match the size difference by standing the mare on a hillside and getting Snoopy on higher ground behind her, but he still could not reach her. Then we dug a ditch, backed the mare's hind legs into it and made Snoopy approach from the dirt pile. Then it happened. The mare was bred, and Snoopy left a few hours later for Texas. I was hoping I did not get an infection from all this maneuvering in the moldy barn, and luckily, I did not.

I was recovering well, and Carolyn started to take me to work, because I was not allowed to drive or carry any weight yet. My energy was coming back, and after I could drive, I went back to work full time. By late spring, I was back on the tennis courts and had started to work on the farm on the weekends; cutting and raking hay and bush hogging. For the first time in a long time, I felt good and energetic. My three-month checkup at the Mayo Clinic was fine. All in all, we had a great summer.

Chapter XXII

Another Health Crisis

By the late fall of 1993, I felt occasional tension in my chest when I was walking uphill or running. I also started to get winded when I was playing tennis. When I went back to the Mayo Clinic in December for my one-year checkup after the liver transplant, I mentioned this to my doctor. He immediately arranged for a brilliant young cardiologist to see me. My heart was tested most of the next day. All was okay, and I was ready to fly home. The doctor recommended one more test—a stress EKG (electrocardiogram). Consequently, I had to postpone my flight. The results were immediately printed out. I had a 97 percent occlusion of my main coronary artery and 85 to 95 percent in the other coronaries. He advised me to stay for cardiac catheterization to see if I needed open heart surgery. I told him that it was only a week until Christmas, that our first grandchild was due right after the New Year, and I had to be there. He let me go provided that I return right after Christmas for the catheterization. He conveyed how serious my condition was by advising me not to carry any luggage, not even a briefcase on my flight.

Upon my return home, I showed the results of my EKG to the family. At this time, our oldest son was already a trained internal medicine specialist, and he was in training for the subspecialty of gastroenterology. He looked at my results and immediately said that I should not fly back to the Mayo Clinic, because it was too dangerous for me to fly.

Christmas was very low key. We wanted to have Christmas for the first time in our new house on the farm. For years, our Christmas tree, always a cedar, came from our farm. We were looking for the right tree when I had a horrible chest pain. I did not feel like walking, so I got back to the house standing on the hitch of the tractor. After I was sitting for a while, the pain eased. Early the next day, we were on our way to the Mayo Clinic. Laz took some vacation time and drove us up there in snow and ice storms.

The heart catheterization was scheduled for six o'clock the next morning at St. Mary's Hospital, a part of the Mayo Clinic complex. From the stress test results, it was quite likely that I needed heart bypass surgery on several of my coronary arteries. The purpose of the catheterization was to determine the exact location of the occlusions that needed to be bypassed. My aortic heart valve was also severely damaged from my endocarditis, and it needed replacement. The feasibility of its replacement could only be determined by a procedure called esophageal echocardiography. This meant that, after I was anesthetized for surgery, a tube with a sensor device would be inserted into my gullet to get a more accurate assessment of the heart valve damage and to determine if it could or should be replaced.

Before the cardiac catheterization started, I told my doctor that if I needed surgery, they could just wheel me into an operating room and do it. He said that they didn't do that. I would need a day or two in between catheterization and surgery. I was awake during catheterization and could see the big obstructing lesions in my coronary arteries; then I passed out. When I came to, I was informed that my coronaries were so obstructed by arteriosclerosis that the split seconds while the contrast material passed through them instead of blood had caused me to go into shock. I was still in shock in the recovery room, and they worked feverishly to get me out of it, because my heart could not tolerate reduced blood pressure resulting in low blood flow through the coronaries much longer. This was even scarier to me, because when I had been in surgery training, I had done research and published a paper on coronary blood flow, so I knew the details already. My doctor then told me my wish would be fulfilled. They would do the operation right away. He said they couldn't keep me out of shock, and the minute the next operating room and heart surgeon was available, my emergency heart surgery would begin.

Shortly after that, a female Pakistani doctor came to my bed and informed me that she would help prepare me for surgery and would be a part of the surgical team. Then Carolyn came in the room scared to death already, because she thought the Pakistani lady was to be my surgeon (she just assisted). On top of that, like all of us, she thought I would have been taken back to our hotel after my catheterization, Instead I was having emergency heart surgery. My mind was hazy, and I had too much discomfort to appreciate the crisis that she and Laz had faced.

We did not have time to worry about this much longer, because the orderlies started to wheel my stretcher to the operating room. We barely had time to say goodbye. I don't know how long the surgery lasted, but when I woke up, the nurse told me they had bypassed four of my coronary arteries and replaced my aortic valve. I asked her if the new valve was a metal one or a pig valve, and she said metal. That was good news, because I knew that pig valves are usually used when a person's life expectancy was five years or less and that metal valves were used if the life expectancy was 10 years or more. As of now, I've had my valve for 17 years, and it is still working.

I never saw my surgeon while I was in intensive care. When I was in a regular room, he came by one Saturday morning with his entourage. I started to introduce myself and thank him for saving my life, but there was no small talk with him. He looked at me and said I was going home the following Wednesday and left. This was the first week of the New Year when the residents rotated, so his resident staff was brand new. My caregiver was a young doctor from Mexico with a chip on his shoulder, but my real caregiver was a physician assistant who made it clear that he was in charge.

After thoracic surgery, a tube is inserted in the patient's chest, which is hooked up to a suction machine in order to suck out any reactive fluid or air from the chest cavity and to expand the lungs. I had inserted and taken out dozens of these while I was in surgery training; however, my chest tube was just not draining properly, and I was getting short of breath. On Sunday morning, when I mentioned this to the resident doctor after a cursory exam, the only thing he said was that he would order a chest X-ray. Sunday afternoon, I was taken to radiology for the X-ray. The heat was turned off for the weekend, and it was freezing cold in there. After the X-ray was taken, I waited and waited in my wheelchair for the orderly to take me back to my room, but he did not come. I could not find any blankets or telephones either, because everything had been locked up, including the exit door from radiology. I was locked in the freezing cold.

Finally, I located a wall phone, but I could not reach it, because I could not get out of my wheelchair. I finally got to the phone, called the operator and told her I was locked in radiology and freezing. Eventually, someone unlocked the door, and an orderly came to get me. When I got back to my room, I asked them to cover me with all the blankets I could have, but it still took me hours to warm up. On Monday, my chest tube was pulled out with great difficulty. Then I saw the reason why it could not drain properly.

It was twisted, and that precluded any suction action. When I pointed this out to the physician assistant, he did not say anything.

Wednesday was my discharge day, and I was still short of breath. My heart was beating up to 120 per minute on average. I did not think I was ready to be released. When the new resident physician came to my room, I told him that, but I was discharged anyway. When he was writing up discharge medication orders, my son came by. The resident mentioned to him that he was supposed to put me on blood-thinning medication, but he wouldn't do it, because I had liver disease. My son told him that I used to have liver disease, but because I had had a liver transplant, I didn't anymore. He urged him to prescribe the regular dose of 5mg per day of blood thinners, because my cardiac bypass and artificial heart valve required it. He did not listen, and instead put me on one-tenth the usual dose, which almost cost me my life.

From the hospital, I was taken to our hotel, where I felt terrible. I called one of my liver doctors, who had examined me before, and he suggested I rest before I flew home. Our itinerary consisted of the following: Rochester to Minneapolis to Louisville. Our flight to Minneapolis was late because of a winter storm, and we barely made the connection. Our plane from Minneapolis had to be de-iced three times, so we were waiting in line for hours. I felt horrible, but I could not get off the plane. If I could have, I would have gone straight back to the hospital. Finally, we landed in Louisville. Instead of gradually getting stronger, I was getting weaker.

During my recovery period, the microbiologist at St. Anthony Hospital was retiring. She was an outstanding professional who had polio and needed two crutches to move around. In 25 years, she had never missed a day of work. She showed up in ice and snowstorms, and she was always on time. I felt that I should attend her retirement luncheon which was held at St. Anthony's. After the retirement party, I ran into my brilliant and caring cardiologist, Dr. Janet Smith. She did not like the way I looked and had an EKG and lab work done. The results were not good, so she admitted me to the hospital for monitoring overnight.

The next morning, I felt better. After I got out of the shower, I developed severe chest pain, and it seemed like I was having a heart attack. I was transferred to intensive care. Luckily, my wonderful and caring cardiologist was in the hospital doing rounds. Dr. Smith asked another cardiologist and a heart surgeon for immediate consultation. My chest pain was unrelenting, and I could not understand why all these doctors were by my bedside. Apparently, I had developed ventricular fibrillation, and my heart had stopped.

In the meantime, little Carolyn, Chris's wife, came to the hospital to bring me some flowers. She went to my original room, but I was gone. The nurse told her I was transferred to cardiac intensive care. While she was trying to find it, she saw the hospital priest walking by and asked him for directions. The priest asked her who she was looking for, and she said Dr. Makk. She then told him she was bringing her father-in-law some flowers. The priest told her she had better take the flowers to the chapel and pray, because something horrible had happened.

I did not know what was happening, but I found out later that the hospital had trouble locating my wife, Carolyn. She was in our garden covering up the rose bushes before an approaching snowstorm and hadn't heard the phone ring. When she came to the hospital to pick me up, she was directed to cardiac intensive care. That was an ominous surprise to her, because she knew all about intensive care.

On her way over, she ran into a surgeon friend of ours. He hugged her and said, "Oh, Carolyn, I am so sorry about Laszlo." When Carolyn asked what he had meant, he told her I had just had a cardiac arrest. My darling Carolyn then took off to intensive care. Outside the intensive care unit, there was a counseling room where relatives were taken for counseling if a loved one had passed away. As she approached, she saw that the rest of the family and the hospital priest were in the room with somber faces. What Carolyn had to go through must have been horrible.

At that time, our youngest son, Andrew, was in Houston working for Enron Corporation when he heard the news. He told his secretary he was leaving for the airport, and by the time he got there, she was to find out the quickest way to fly to Louisville and secure tickets. Andrew arrived at the hospital within four hours after he had left his office.

When Carolyn was brought in to see me, the doctors told her that they were trying to stabilize me so that I could be transferred to Audubon Hospital, the local hospital that specialized in heart surgery. In the meantime, the news went through the hospital.

Finally, I was stable enough to be transferred to Audubon. As I was wheeled away on the gurney, dozens of people lined the corridor, from the administrator on down to orderlies, to say goodbye. I saw many wet eyes, including those of my laboratory staff standing nearest to the elevator. I threw them a big kiss as the elevator door was closing. Then I saw some of them smiling.

My cardiologist, Dr. Janet Smith, and my heart surgeon rode in the ambulance with me. The doctor ordered blood earlier and wanted to have it next to the ambulance. My family closely followed in our Jeep with the blood in their laps. When we got to the hospital, more doctors started to work on me. I also had signs of a cardiac tamponade, which was an accumulation of blood or fluid around the heart that had compressed it, thus decreasing its pumping function.

It was decided to put me through cardiac catheterization first. There was a large fibrin clot in my main coronary artery also occluding the bypass vein. They tried to dissolve the clot. After some maneuvering, they were successful. The fluid around my heart was not blood but an accumulation of postcardiotomy fluid, which was usually absorbed in time and can be drained later if necessary. My pain became less, too.

While I was in the catheterization room, one of the heart surgeons got a call from India. The caller told him to do whatever it took to get me through and to see to it that I got the best doctors and care. He also assured him that the cost would not be a problem. The surgeon came in and said, "Laszlo, I thought you were a poor refugee, and I just got a call from India from a friend of yours, some executive who told me I better get you well or else." The caller turned out to be Joe Sutton, a dear friend and Enron executive. Apparently, Andrew's secretary had told Joe's secretary about my crisis, and she called Joe in India to inform him of the latest development.

Next, the leading heart surgeon in Louisville, Dr. Alan Lansing, stuck his head in the cath lab, He said he was supposed to go to Milwaukee that night to give a speech but had cancelled it in case I had needed him. That was very reassuring. I was transferred to the cardiac intensive care unit, where I started to go downhill again after a couple of hours. I was moved to another room in intensive care, and eventually, I went to sleep. When I woke up and became more aware of my surroundings, I noticed the room was quite large and that there were many electrical outlets in the walls. I asked the nurse what kind of room it was, and she said that it was where they kept the sickest patients, such as artificial heart recipients. Slowly, I started to recover.

I would have never made it through if I hadn't gone to the retirement luncheon and ran into my cardiologist—my dedicated cardiologist who never left my bedside for eight hours. Today, I am grateful to the many doctors who came by to see me even in the worst ice and snowstorms.

This had all happened because one of my four bypasses had been connected improperly at a right angle instead of a smaller angle, and because I had been given an inadequate dosage of blood-thinning medication ordered by the resident physician at the Mayo Clinic (all of which almost resulted in my death).

After I pulled through this crisis, I started to recover nicely. Pretty soon, I went back at work, visited the tennis courts, and started to work on the farm again. I never realized how sick I had been and what a close call I had experienced.

I am now eternally grateful to my loving family for their support, and I am sorry that I had to put them through that. I am also thankful for my wonderful and brilliant physicians, including our two doctor sons. After my recovery, I had more energy and more strength, and pretty soon, I felt like I had in the good old days before my liver problems had started. Everything was going well for a while, and then my hospital started to go downhill.

Chapter XXIII

The End of St. Anthony's

St. Anthony Hospital was in the process of major expansion when I arrived in 1967. For years, the administrator had been a delicate, small nun named Sister Francis Anne. She had to fight for the expansion at that time because the Motherhouse had been fiscally conservative and had not wanted to spend any money. She eventually told me the reason that the Motherhouse finally relented was because she threatened to leave the order if they continued to keep her from modernization and expansion.

At that time, the St. Anthony's Pathology Department and Laboratory was in disarray and not trusted by the medical staff. The hospital's pathologist was not board certified and the doctors on the staff did not trust his diagnostic work. Sister Francis Anne and I went on a tour of the facility, and at one point, she grabbed my arm and said, "Doctor, I want you to build the best lab there is."

She pointed out that they now owned the whole city block across from the hospital. When I asked her what they had planned to do with it, she said maybe they would turn it into a parking lot.

I suggested they consider building a large doctor's office building, because in Houston, the only community hospitals that did well all had an adjoining large medical office building. She liked the idea. I even mentioned that if they charged no rent for the first six months or half rent for a year to young doctors just starting their practices, the office building would be full with the brightest doctors, and they would fill the hospital with patients. She was very excited about the idea, but nothing happened until 10 years later when they built a very small office building. They leased part of it to the hospital's architectural firm, which of course, could never admit any patients to the hospital.

The hospital offered basic radiotherapy services. I learned that the renowned head of radiotherapy at the University of Louisville School of Medicine was leaving. He had a beautiful historical home, and I knew he was reluctant to leave it. I thought he might consider a local offer, even though he had already received an offer to be the Chairman of Radiotherapy at Indiana University.

I went to the administrator and suggested St. Anthony's recruit him to run its radiotherapy department. She asked what it would take to get him. I told her that he wanted to build a radiation therapy center. We went to the board with the idea, and they enthusiastically supported it. The Board Chairman and another senior board member traveled to the Motherhouse with the administrator to add weight to the request, but the Motherhouse inevitably turned it down.

After this, the Board Chairman told me we would just have to keep putting the proposal in front of the Sisters until they approved it. Finally, they approved it, and the hospital was able to recruit the brilliant radiotherapist and his associates. Pretty soon, the new center gave more radiation treatments than any other radiation center between Chicago and Houston. It brought prestige to and generated much profit for the hospital.

When we needed a new autoanalyzer for the lab, I worked out a very advantageous lease arrangement with the manufacturing company, because I had previously published research on their instrument. The hospital's purchasing director informed me that such arrangements were usually not allowed. The hospital's philosophy was the following: Earn the money first and pay cash.

I was also ready to make an arrangement with a laboratory supply company that involved having them come once a month to replenish what we had used, and then bill us for it. Unfortunately, we had to send all requests to the purchasing department. They did the actual purchasing. The big difference was that we had to pay for our supplies ahead of time with the way purchasing handled it; whereas doing things my way, we only paid for the supplies we actually used after the fact.

Sister Francis Anne, the original administrator, was getting too old and eventually decided to retire. There was no suitable sister in the order to replace her, so they sought out the first layperson to be placed in this position. They hired Assistant Administrator Mike Abell who had a Masters in Hospital Administration. He was an excellent administrator, energetic, and young. He, too, soon became frustrated with the Motherhouse's conservative policies. The Sisters who were members of the governing board wouldn't loosen their grip on their hospitals. No wonder their hospitals couldn't exist in a modern, competitive healthcare environment.

Mr. Abell was terrific. He prudently observed that the population was moving to the suburbs and wanted to build a hospital out in that area. He located a large parcel of land that belonged to the Catholic Archdiocese in a perfect location. A seminary was once located on the property. The seminary building with a large kitchen and laundry still remained. Abell succeeded in getting the Archbishop to agree to sell us 50 acres with interest payments to be made only for the first ten years. It was ideal. Seventy percent of doctors in Louisville lived within a mile and a half radius from the site. Unbelievably, the Motherhouse just said no.

Our neighbor, Baptist Hospital, saw into the future and built a hospital in the east end of the city, eventually closing the original one. The hospital at the new location had two adjoining doctor's office buildings. Baptist Hospital East was full in no time. They decided to add a radiotherapy center, and our radiotherapists were invited to run it. Our hospital was offered a 50 percent share in it, but the Motherhouse once again said no.

Finally, Mr. Abell could not take the frustration anymore, and left to become the administrator in a hospital three times larger than St. Anthony's.

Things went straight downhill from there. However, we kept our focus on meeting our patients' needs. Once a year, we surveyed the medical staff to evaluate our services. Every year that the survey was done, we were judged to be the best hospital laboratory in Louisville. Despite our outstanding radiology, radiation oncology, and laboratory services, our hospital census kept slipping. Doctors felt that they were more welcome at other hospitals, and that our institution had become too bureaucratic.

For instance, I had an extremely wealthy friend in Florida. His household staff consisted of 17 employees. He had a complication from disc surgery and was looking for some medical help. He was very independent and did not believe in government handouts; therefore, he did not have Medicare. I suggested that he come up to our hospital, and I organized a team consisting of the best internist, orthopedic surgeon and neurosurgeon to evaluate him.

On the day of his scheduled arrival, when he was already in the air flying to Louisville, the head of our admissions office told me that my friend could not be admitted to the hospital because he was 85 years old and had no Medicare. Anyone past the age of 65 could not be admitted unless they had Medicare.

I explained his financial situation, and that he would pay his bill in full and likely give the hospital a generous donation. It did not matter; policy was policy. Finally, we agreed that the Director of Guest Relations would greet Mr. Smith at the front of the hospital and take him to his room. The medical staff would take over from there. He was greeted but told that nobody could be admitted to the hospital without going through the admissions office. That was the last thing Mr. Smith needed after his long trip. We all agreed that this was an emergency. He would be taken to his room where an admissions clerk would meet him to formally admit him. He would not be bothered with Medicare processing or asked to pay a deposit. (They wanted a large deposit, particularly because he did not have Medicare. I told them I would guarantee the payment of his bills.) The orthopedic surgeon told Mr. Smith that there wouldn't be a bill from him. Known for his generosity, Mr. Smith had given a big donation in his name to Duke University, where he had been trained. It was a typical St. Anthony's case. All hospitals would love to have a wealthy patient like Mr. Smith, but at St. Anthony's, I had to beg and plea with clerks who had no understanding of his situation and almost hindered his coming to the hospital.

St. Anthony's continued to slide. downhill. The ineffective administrator that replaced Mr. Abell was finally terminated. This time, the Sisters from the Motherhouse arranged for our local board to interview new candidates. The one we recommended was okay at the beginning. He called himself "CEO" instead of Administrator. As the hospital revenue started to decrease, they were terminating more and more direct caregiving employees, but not the bureaucrats. The lab was no exception. At first, our department was moved from reporting to the Administrator to reporting to a new Assistant Administrator.

Unfortunately, neither of them knew anything about a medical laboratory. For instance, once the Assistant Administrator and I were standing next to our big autoanalyzer, He said he thought we should put one into our budget. I pointed out very carefully that we had owned one for a long time, and as a matter of fact, we were standing right next to it. He was getting a high salary to oversee the lab when he did not have the slightest idea what a lab was. More and more doctors took their patients to competing hospitals where they felt more welcome.

This new administrator was also a great believer in management consultants. These consultants took up our time interviewing us and kept us from doing our work. It didn't take us long to figure out that they always seemed to recommend what the administrator wanted to do. So again, non-caregiving personnel were hired, and the number of direct caregivers was reduced. There were more nurses carrying clipboards and wearing street clothes than nurses in uniforms taking care of the patients.

The Assistant Administrator, to whom the lab reported, would come by and demand employee reductions. One day, he took a copy of my contract with the hospital out of his briefcase and said, "I carry your contract all the time, and from now on, we will review it every three months."

I told him to let me know if they were concerned about me earning my keep. That took care of it. He never mentioned my contract again. In fact, he became very respectful and complimentary.

On another visit, he closed my door and informed me that one of my junior partners had approached the administration and had said he could do my job for less pay than what I received. The administrator assured him that they were extremely pleased with my services and felt fortunate to have a pathologist like me. I had been good to this junior partner, and just could not take his stabbing me in the back.

I asked for my attorney's advice, and he told me to fire him. I also asked a friend, the hospital's malpractice attorney, and he said the same thing. I fired him that day, giving him 90 days to find another position. He was a talented and smart pathologist, but he had been a consistently critical whiner from the start. When they he first came to town, we had invited him and his wife for dinner. They showed up several hours late with his parents, children, and their maid. We suddenly had to serve seven people instead of four. Then they got upset when we seated their maid at the table. He was very cynical and critical of my other partner, the medical staff, and of the hospital in general..

In 1995, between Christmas and New Year's, I got a call from our hospital attorney who was also secretary of our board. He wanted to know if I had heard anything about the hospital being sold. I told him I had not, but I would keep my ears open. If I found out anything, I would be in touch. We did not have to wait very long. The next morning, out of the blue, a department head meeting was called. At the meeting, a Sister from the Motherhouse informed us that the hospital was being sold to Vencor Corporation, which operated subacute care hospitals and nursing homes.

We all felt betrayed. The Sisters did this without the board members' knowledge. What was even worse was that a large group of doctors had approached the Motherhouse earlier to purchase the hospital and operate it with observance of Catholic-based hospital rules and regulations. Their offer was turned down, even though their offer had been more than five times that of Vencor's purchase price. I heard the sale price was $5 million. The land which the hospital occupied was worth more than twice as much. This was the tragic end of a wonderful Catholic hospital and the termination of hundreds of loyal employees.

Chapter XXIV

The Vencor Year

The morning after the announcement of the sale, the president or chief operating officer of Vencor met with medical directors and medical staff members. He announced that they would try to keep as many employees as possible, welcomed the medical staff, and asked for ideas to improve services. He asked me to stay on with the new Vencor hospital. He assured me that my contract would be basically the same as I had had with St. Anthony's, and it would be rewritten with Vencor in the near future.

In March, the operations were transferred to Vencor. We had the last mass in the chapel, for which a lot of Vencor executives showed up.

I was assigned a Vencor executive to be my contact man at corporate headquarters. I noticed that their 42 hospitals and nearly 300 nursing homes had either no laboratory or one with a very limited capacity. Nearly all of their lab work was sent out to various commercial laboratories. We had the capacity, skilled people, and room to set up a reference laboratory; whereby all lab work could be sent to us, and we could perform the tests promptly and transmit the results immediately. With the right instrumentation, we could have the results back, even to the West Coast, faster than the commercial laboratories. The results would also be uniform.

I designed a business plan and presented it to the president and other executives of Vencor. They were very excited about it and authorized me to proceed. They even sent me to other Vencor hospitals to have firsthand experience of what their lab operations looked like. I was really excited that now we would be able to give high-quality, uniform service to their patients across the country. My old motto—in every crisis there is a seed of opportunity—was proven to be true again.

In their subacute hospitals, many of the patients were on respirators and required a lot of microbiology services. They suggested that I implement a microbiology lab first. We quickly obtained cutting-edge instrumentation, and our trained personnel were ready to roll. I had an interstate laboratory license and transferred it to Vencor for out-of-state work.

Then, I was asked to wait with the implementation of other lab work. In the meantime, I still didn't have a contract and hadn't been reimbursed for my work or my associate's work. Every time we talked about it, they assured me that it was coming, and that they were still preparing it. This was going on too long. After we went through it one more time, I told them I would have what we agreed to typed up and forwarded for their signature. They agreed. I sent the contract, but there was no reply. Finally, I received a longhand note written on personal stationery from my contact executive informing me that the contract could not be signed because he didn't recall any of the particulars we had discussed. He was in the hospital three or four times a week and usually stopped by to visit and discuss details, and had previously agreed to the particulars of the contract.

Prior to this, a new laboratory supervisor appeared, apparently sent by Vencor headquarters. She behaved oddly, and when I inquired about her training and experience, she got very hazy and never gave a specific answer to my questions. The hiring and firing of employees in the lab was under my authority; however, she started to fire some of the most experienced people behind my back, and when I questioned her about it, she gave vague answers. Along with the dismissal of talented and dependable employees went my pride and decades of effort to have the best and most qualified people in my department. When I protested to my Vencor executive contact, he said they would look into it and give me a follow-up answer, which never happened, just like setting up the rest of the lab never happened.

I got very busy making phone calls to help the laboratory employees who had been terminated find new jobs. At least that effort on my part was very rewarding and successful.

We started getting interesting bacteriology cases, some of which could be publishable. I suggested to Vencor that we host a scientific symposium once a year featuring a nationally renowned speaker and a black-tie dinner. This type of event would bring prestige to Vencor and attract doctors. They were very enthusiastic, but that was as far as that idea went.

It had now been nearly seven months since Vencor had taken over, and I had neither a contract nor payment. I decided to seek the advice of my friend David Jones (CEO of Humana). He was surprised by my problem, and personally called the Vencor CEO whom he knew well. He explained to him that I was a highly respected laboratory diagnostic pathologist in the medical community. He also mentioned that a leading surgeon had told him that whenever other pathologists had trouble figuring out a difficult case, they would come to me for help. The Vencor CEO thanked him, and said he would take care of it. Shortly after this phone call, I was asked to go see my Vencor contact at corporate headquarters. He was sorry that finalizing my contract had dragged on so long. Out of the blue, he offered a contract proposal, however, with one-third of the compensation of what we had agreed upon several times in past meetings.

Suddenly, it hit me that they might have been trying to get rid of me, hoping I would walk out after I had heard this insulting offer. I told him I would think about it, but he insisted I had to decide whether or not to accept their offer on the spot. Finally, he said that we had better part company and that my last working day would be December 31st. I demanded that they pay me the sum of money that we had previously agreed upon in our verbal contract up to that date. He agreed to that.

Later, I received a message that said I could pick up my paycheck at the comptroller's office on December 31st at 5:00 p.m.

Customarily, the hospital had a farewell party for those who were leaving, but there was not one for me. My staff walked by my office and waved goodbye, and at times, they wiped tears from their eyes. I could sense they were worried about their job security if they were seen talking to me.

On December 31st, I went to pick up my paycheck on the way out of the hospital. It was a lonely walk, and not a single person said goodbye. This was the end of a brilliant career in the hospital where I had built a renowned laboratory which received every conceivable award and recognition, and had transformed the lab for Vencor patients' specific needs.

My involvement in Vencor was not over yet. On New Year's Eve, I got a frantic call from the chief engineer of the hospital. He informed me that he had been ordered to throw out all diagnostic microscopic slides and pathology reports. These slides and reports had to be legally kept for at least 25 years. He tried to explain this to Vencor, but no one would listen. This was a horrible decision, particularly concerning cancer patients. If they needed a new operation, it was imperative to review slides from previous operations to see if the lesions were interconnected. Plus, we were talking about three or four tons of slides and records.

I told him we needed to get a truck and maybe we could store them on our farm, and if we had to, we would provide custodial services. He said he knew a medical record storage warehouse, and he would try to get to them that night. But he had to give them a billing address and asked if he could give them my home address. I started to get monthly bills from the storage company for storing the hospitals microscopic slides and records. After a while, I contacted the Motherhouse, requesting them to take over the payments for their stored material and reimburse me for previous payments. They agreed to do so, and I eventually received my payment.

After I left Vencor, I still had my small private lab to practice in. It was a lot of fun—no more meetings, no dealing with paranoid, lying supervisors and administrative personnel. I just focused on practicing good laboratory medicine. I was fortunate that many of the best doctors referred their lab and pathology testing to me. One of these referring doctors was an internationally renowned cancer surgeon. He had a lot of referrals from other surgeons for mostly complicated cancer cases. He routinely requested that they send me microscopic slides and pathology reports before he would see the patients so that I would have time to study the case and forward him my consultation report. This was quite an honor. He was also a famous malpractice expert witness for the defense. Again, he had the slides sent to me first and relied on my reports. This meant depositions and occasional court appearances. He was also working on a cancer book and asked me to write the chapter on diagnostic and prognostic tumor markers or indicators.

I gave a presentation on the use of tumor markers at the European Cancer Research Congress; however, there had been so much progress made and so many new types of tests developed that I became a regular at the medical school library doing research. Unfortunately, my friend had two strokes after heart surgery and then passed away. That was the end of his book.

Right after Vencor, two prominent pathologists invited me to join their practice, which was a non-hospital-affiliated private laboratory. I was glad to join. We would consult with one another on difficult cases. I was also free to travel, because they just took over for me whenever I needed or wanted them to.

A nurse physician's assistant also invited me to set up their laboratory and supervise its construction at an infertility physician's office. I did that, and I also arranged for them to get the equipment free and pay only for reagents used. I had a wonderful, relaxed, high-quality practice.

When it came time for me to retire, I just turned over my practice to my two colleagues. A lot of pathologists sold their practices, but I had always felt a pathologist got his referrals because the clinician had faith in his diagnostic ability. I considered that an honor and not a saleable commodity. I was very happy to leave my practice in good hands, without financial considerations.

Chapter XXV

Carolyn's Cancer and Retirement

In 1996, the family was planning to celebrate Easter together at our Palm Beach home. Carolyn had gone down early. In a single day, our lives changed.

The day was not going well. This particular day, I stopped by our farm to check on things before I left town to join my wife in Florida. Sadly, I ran over Carolyn's favorite dog and killed it. I dreaded telling her. Then, upon my return to town, I heard on the news that a plane carrying an American business delegation had crashed in Croatia and that everyone on board had perished. I was alarmed, because I knew my best friend, Joe Sutton, had been on the plane. His name had been on the flight manifest. Very upset, I called and found out that Joe had luckily missed the flight and was now flying on another chartered jet. He landed only to find the delegation had not and sadly would never be arriving at their destination had crashed into a mountain. We were relieved that God's grace and good luck saved Joe from the fate of his cohorts.

That same afternoon, I spoke to Carolyn in Florida. She told me that while she had been showering that morning, she had felt a lump in her breast. She went to the doctor immediately to get a mammogram, and the radiologist now needed to speak with me. I felt nauseated when he told me the lump was most likely cancerous. We decided to take Carolyn home to Louisville right away to pursue a biopsy and a potential mastectomy.

The biopsy did prove that it was cancer. From there, things moved quickly. She underwent a total removal of the breast and auxiliary lymph nodes. It was good that none of the lymph nodes or the remaining breast showed residual cancer. Now it was time to fight the cancer with chemotherapy. Thankfully, she responded very well to chemotherapy. After a waiting period, it was decided to go the way of caution and have her remaining breast removed as well. The final stage of this initial cancer episode was to undergo reconstruction. She opted for implants and recovered very nicely. She was blessed with survival. We could once again focus on the "normal" things in life, and we enjoyed our family and friends more than ever before.

I had been contemplating my retirement for some time. I was looking forward to having a more active role in our farm and spending more time with Carolyn and our grandchildren. Then once again, everything practically changed overnight.

It was the week after Memorial Day in 2004. We were in our home in Florida with our eldest son, Laz, his wife, and their young children. My dear Carolyn started feeling really tired and seeing double and developed unsteady balance. Laz, our ever-duty-bound son, and his family were flying back on Friday afternoon, because he was on call for the weekend. Most doctors would just trade the weekend with another member of their group to add a weekend to their vacation, but not him. Duty came first. Shortly after his arrival, we got a frantic phone call from him. He said he had discussed Carolyn's case with a neurologist who thought she had an acute hydrocephalus and needed emergency medical help.

I immediately started to consider what would be best for her. We had two choices: either go to an emergency room here or try to fly home, where our family had many good medical connections. In Florida, we would be in a strange medical environment, not knowing most of the doctors or hospitals, not knowing how to get the best instead taking the one who was on call, who may or may not be the best choice. In Kentucky, we knew all the best doctors, and Laz, Steve, and I were known and respected among the doctors and hospitals there.

My choice was to take the first flight to Louisville. If we couldn't get on a flight soon enough or if they wouldn't let us board, we would then go to the emergency room in Florida. I was concerned that the airline wouldn't let us get on the plane, because Carolyn's unsteady gait could have been misconstrued as drunkenness, in which case they could have prevented us from boarding. In every crisis in our life, the Lord reached us and paved our way.

We got seats on a Delta flight that was leaving in an hour. We had left in such a hurry to get to the airport that we left the kitchen door open. When a friend came to check on our house, he found the kitchen door open and thought we had been burglarized at first, but found everything undisturbed.

Fortunately, we got to the airport in time, and as we proceeded thought the first picture ID check, I kept talking to the checker order to keep his attention away from her unsteadiness while I supported Carolyn by hanging on to her belt in. We then made it through the second picture ID. I don't know how she did it, but she managed to walk through the detection booth on her own. Finally, we were on the plane. She was holding up okay, but in the Atlanta airport, when we changed planes, she lost her driver's license. Luckily, we didn't have to go through a picture ID check by that time.

In Louisville, Laz picked us up at the airport and drove us straight to the Baptist Hospital East emergency room, where he had prearranged everything. We had landed in Louisville at 11:15 p.m., and by midnight, we had the CAT scan, which showed a large tumor in the pineal body at the base of her brain. A neurosurgeon was already examining her by 1:00 a.m. He told us Carolyn's cerebrospinal fluid circulation had been obstructed by a large tumor that produced high pressure on the brain. She was suffering from hydrocephalus (fluid build up on the brain) and needed a tube called a shunt placed into her brain The tube would be pulled through under the skin of her scalp, neck and chest wall to her abdominal cavity. It would drain the accumulated cerebral fluid and relieve the excessive pressure on her brain.

She couldn't be taken to the operating room yet because she first needed to be stabilized with medications, which took a day and a half. The surgery went well. After she recovered from anesthesia, she felt better right away. She then had some more MRIs done. The brilliant radiologists and neuroradiologists carefully studied the results and concluded that it was most likely a cancerous tumor, which may have also invaded surrounding vital structures. When our neurosurgeon studied the results, he felt an operation may have been her only chance but pointed out that such an operation carried major risks. Thirty percent or more of such patients would wake up blind or paralyzed, and some didn't wake up at all, remained in a coma or worse. When Carolyn asked him how many cases he had done, he said he had performed only a few.

With her history of breast cancer, the most likely diagnosis would be recurrence and spreading of the cancer to the brain. I kept wondering if the breast cancer had spread to other organs, too. As she was being discharged from the hospital, I asked the neurosurgeon if she could have a new test called a PET scan, which could detect cancers throughout the body. He ordered one right away. The results indicated tumor presence near the windpipe and possibly in the right lung. The bad news wouldn't stop coming at us.

As I was trying to decide what to do, I suddenly remembered the wife of a dear friend, Rick Park, once had a noncancerous brain tumor near the location of Carolyn's tumor. A neurosurgeon had referred her to the Mayfield Clinic in Cincinnati, Ohio, where she underwent a successful operation. After I looked up Mayfield Clinic, I was very impressed that they had more than 20 neurosurgeons, 11 doctoral-level neuroscientists, nine neuroradiologists, and two or three neuropathologists. I also found out that the heart and soul of Mayfield Clinic was the world-renowned Dr. John Tew. His practice focused most of its efforts on neurosurgical cases involving the base of the brain, and therefore, Dr. Tew was an expert in this type of surgery. After I had checked out other possible neurosurgeons, I knew he was our best hope.

We arranged an appointment with Dr. Tew for the following week. While we were waiting, our son Steve suggested that we have a mediastinoscopy done on Carolyn to evaluate if there was a tumor around the windpipe and lung area as the PET scan had suggested. Mediastinoscopy was an operation whereby the surgeon passes a tube down beneath the breastbone toward the heart after the patient was anesthetized, looked around, and removed some tissues. A frozen section showed no tumors. The next day, I got a call from the pathologist. They had found small groups of cancer cells in densely fibrous lymph glands. This was really bad news. She probably had had an undetectable tumor at that location at the time of her first breast cancer surgery 10 years earlier. This study suggested that she had had cancer at that location then, and although the chemotherapy killed most of the cancerous cells at that time, fibrous tissues had replaced them, and a few mutations had survived.

After some additional studies, Carolyn's case was presented at the Mayfield Clinic neurosurgery conference. There was considerable discussion, because even in this large center, they had never dealt with a patient that had breast cancer spread to the pineal body of the brain.

Dr. Tew recommended surgery. Because most neurosurgeons had very limited experience with pineal body tumors, Carolyn asked him how many patients with pineal body tumors he had operated on. He said more than five hundred children—noncancerous pineal body tumors were mostly found in children—and 37 adults. This was more than anyone else. When Carolyn inquired further about how many had survived, he said all of them. Then Carolyn wanted to know how many had major complications. Dr. Tew said one patient had suffered some paralysis of her left arm but that he saw her every day, because she was his neighbor. We felt God had led us to find Dr. Tew, so we scheduled the surgery for the following week.

The night before we were due to leave, a neurosurgeon friend called me and advised, "Don't do it."

I tried to find out what had prompted him to say this and asked how much experience he had with pineal body tumors. He said he had only worked on one case and that the patient had died. I decided we would proceed with the surgery despite this advice and despite the presence of cancer elsewhere. Never give up fighting a disease.

Our whole extended family arrived in Cincinnati the day before. Around two or three o'clock in the morning, I got a call from Dr. Tew's nurse, who informed me that Carolyn's surgery had been canceled because Dr. Tew had an out-of-town emergency. I later found out that his mother had died. The nurse told us to go home, and they would be in touch.

A few days later, she called saying Carolyn's operation had been rescheduled. During those days, I never had any doubt that Carolyn would make it. After we met Dr. Tew, his humble, loving, and caring personality made us feel like we were in God's hands. Dr. Tew told us the operation would last about five hours. Our entire family was together again, and we sweated it out together. The five hours passed, and there was no word from Dr. Tew. Then six, seven, eight, and nine hours passed, and there was still no word. With each hour passing, I got more and more worried. With my surgery and pathology training, I knew that each hour delay may mean bad news, and so did everyone else in the family. We tried our best to comfort each other and keep hope that she would be okay.

After nine and a half hours, Dr. Tew came out to the waiting room and said, "Dr. Makk, please come with me." As a pathologist, I knew this could mean the worst outcome. At this point, I didn't know whether Carolyn had or hadn't survived the operation, and Dr. Tew may have wanted me to know the new first if she hadn't. Thank goodness he had a very brisk walk and it didn't leave me much time to worry. I followed him right into the operating room where Carolyn was laying on the table, and I saw that she was breathing. Then Dr. Tew said to her, "Carolyn, move your right leg, left leg, right arm, left arm, smile," and Carolyn did it all. This was the most joyous moment of my life. She had not only survived the critical surgery but hadn't been paralyzed, blinded, or comatosed. Now, she was waking up! Then I kissed her, and she tried to smile. Life was great.

Then Dr. Tew informed us that he had been able to remove about 80 percent of the cancer, because the rest was extending into nearby vital areas and an attempt to remove it would have resulted in catastrophic consequences. He recommended that she undergo the gamma knife (a knifeless surgery), a form of super radiation therapy, to eliminate the remainder of the cancer in her brain after she had recovered from the operation. He also recommended chemotherapy. She recovered nicely and never complained. The chemotherapy was started before she left the hospital and given under the care of Dr. Elyse Lower, who is a renowned oncologist who specialized in advanced stage IV breast cancer treatment. Stage IV breast cancer was one that spread (metastasized) from the breast to distant locations in the body. Dr. Lower was also the author of pioneering research and publications on the medical treatment of breast tumors that had metastasized to the brain.

Next, Carolyn was transferred for rehabilitation to Baptist Hospital East in Louisville. After dismissal from there, she received rehabilitation service on an outpatient basis.

Once she was stabilized, it was time to go back to Cincinnati for the gamma knife procedure. Dr. Warnick, chairman of the Mayfield Clinic and the chief of radiation oncology, administered it together. I had had medical training in good places, and I knew good doctors; however, I was overwhelmed by the dedication and brilliance of our doctors.

The in- and out-patient rehabilitation at Louisville's Baptist Hospital was supplemented at home. My physical therapy training came in very handy. She was a wonderful patient and never complained. Carolyn progressed so well that she was able to walk on her own in no time at all, and not long after that, we attended a dinner dance. She then received chemotherapy in Cincinnati regularly for five years. Thank God her hair and her pleasant demeanor had not been lost after chemotherapy.

Now that Carolyn was well, it was someone else's turn to get sick in the family. Once again, it was me. I was enjoying a very active life when I started to develop a tightening in my chest, particularly when I was walking up steps. Cardiac catheterization revealed two new blockages in my coronary arteries, one lodged in a very dangerous location. Two stents had already been placed in my coronary arteries just one year after my open heart surgery.

After Dr. Bill Dillon, a young and very bright invasive cardiologist, discovered the challenging location of the new blockages, he consulted his father, who was a well-respected invasive cardiologist out of Indianapolis. All of the risks were explained to me, and it was my decision whether to proceed with the high-risk and complicated stent placement or not. I decided to go with it. During the procedure, pieces of plaque were dislodged. I developed severe chest pain going to my arms, and I went into shock. Suddenly, there were additional cardiologists and a heart surgeon in the catheterization room working in feverish activity. Finally, my chest pains started to ease, and I found I could breathe better. The procedure had been finished, and I had to give credit to the doctors' courage, knowledge, and guts. They got me through, and I was good and fit again; however, I developed kidney and heart failure soon after that. My great doctors pulled me out of heart failure then and succeeded in getting my kidney failure under control as well.

There were happy days again, and everybody was in good health in our growing family. Our delightful grandchildren were (and for me continue to be) a constant source of happiness and surprises. Each of our grandchildren has been a special gift from God.

After Carolyn's brain surgery, I decided to retire so that I could take care of her, and it was time for it anyway. We had the time of our lives. Though I had started out as a penniless refugee immigrant, we now had a great American family that now included a third generation to enjoy Makk family traditions.

We would gather at our home in Louisville and spent much time together at the farm. Carolyn had a special talent for decorating and selecting antiques for our homes—and for making them warm and welcoming, We also had a house in Palm Beach, Florida which we visited regularly. It was nice to see the family enjoy their vacations in the Florida home.

Our life was very happy, and we never gave up hope, regardless of what challenges we faced in life. We remained optimistic even during our catastrophic illnesses.

However, the family didn't stay in good health for very long. Our oldest son Laz, who was a dedicated doctor, had suffered from ulcerative colitis for some time. Now it was turning into cancer, which was usually very aggressive with ulcerative colitis. His first operation failed, and then he had to undergo a very major operation at Cleveland Clinic, which lasted seven hours. He had a major postsurgical complication; however, he eventually recovered. He has a wonderful wife and three young children. He is well today with a very busy medical practice. Once again, we were all in relatively good health. Carolyn and I enjoyed our busy retirement. We loved Palm Beach in the winter and Louisville and our farm in the summer.

Carolyn was doing well in 2008. She was in the fifth year of continuous chemotherapy. Then in the spring of 2009, she developed some hoarseness and started to lose her voice. Medical workups revealed a paralyzed left vocal cord caused by a new spread from her breast cancer into the upper neck, adhering to her jugular vein and carotid artery. Dr. John Tew offered to operate on her, but could remove only a small part of the tumor. He recommended radiation therapy to control it, which she received every day for six and a half weeks. This was not good news, but my saying proved true again—somebody was always worse off than you were. Follow-up studies indicated that her residual tumor had shrunk, which was a sign of successful treatment. It also indicated that future reoccurrences would also likely respond to radiation treatment. Our joy didn't last very long as she developed mild swelling and droopiness of her left eyelid. She was watched very closely. We would have another medical challenge at hand with a new recurrence if this symptom did not disappear.

Chapter XXVI

The Final Challenge

After this most recent round of radiation, Carolyn was still very weak and made every effort to regain her energy. She was still losing weight at an accelerated rate. This alarmed her oncologist and all of us. We made a big effort to increase her appetite and tried to get her to gain some weight. Laz did some research on intensive care nutrition, and we used all his expertise. Steve also knew about weight problems, and Chris tried to figure out what she might have wanted to eat and cooked it. Andrew was sending the most delicious care packages. Our daughters-in-law frequently showed up with food.

I cooked everything that she liked, from the most delicious steaks to Hungarian food. I tried very hard to cook what she might have liked, and she tried very hard to eat it; however, it was to no avail. She was still losing weight at an accelerated rate. She never gave up hope or complained, and she always made her best effort to eat. Besides her weight loss, her energy was gradually being sapped as well.

We were planning a big Christmas with the entire family coming for dinner. Andrew flew home from New York four weeks early for a visit, because he and his wife were expecting their first child in January and they did not want to get bumped or contract the flu on the plane. As he was getting ready to leave, a strange feeling took over me. Everybody was so loving and healthy. Could this be real, and could it last? Suddenly, I had my doubts.

After Andrew left, Carolyn was getting weaker, and for this reason, we did not attend Christmas dances and parties. We agreed that I would cook the main dish, a nice tenderloin, and the kids would bring side dishes. We did not do our traditional Yorkshire pudding, because I did not want Carolyn to wear herself out. She was the only one who could cook it perfectly.

After Christmas, she grew even more tired. On December 30th, we went to Cincinnati for chemotherapy just before leaving for Florida. Her oncologist, Dr. Lower, was a great doctor with a wonderful personality. But this time, she was sterner and very forcefully instructed Carolyn to gain weight, because after we got back from Florida, she would have to receive stronger chemotherapy. This would be absolutely necessary, but they could not do it if she didn't gain weight. Weight loss was a frequent side effect of this treatment, and she could not tolerate it unless she gained weight before receiving it. After we got back to Louisville, she tried her best to eat, but she had no appetite.

Then we headed to Florida. Carolyn was always happy to come to Palm Beach. It seemed her energy level and appetite improved. We both loved Too Jay's takeout. After a day or two, her voice got weaker, and she seemed to cough more. I suggested she get medical attention, but each time, after she finished coughing, she said she felt better. She had coughed like this before, so I took her word for it. She ate a good dinner that evening. While I was on the phone with Laz, he overheard Carolyn's coughing. He remarked that it didn't sound good.

After we went to bed, we watched television and planned a trip to New York, by way of Washington, D.C., with our grandchildren. It would coincide with the christening of our yet-to-be-born granddaughter, Hannah Carolyn. Then she started coughing more. I checked her vital signs and thought everything was all right except for increased respiration, up to 24/minute, which indicated mild lung distress, but was stable. After she went to the bathroom, she looked even worse, and her respiratory rate jumped to 34/minute which indicated sever lung distress. I immediately called EMS. I gave her a Heimlich maneuver in case she was choking, which was sometimes a complication of vocal cord paralysis.

While I was on the phone to EMS, she fell back from a sitting position. She was not breathing, and she did not have a heartbeat. The EMS was prompt and did a very good job with resuscitation, but their efforts were in vain. She was promptly transported to Good Samaritan Hospital's emergency room, but all they could do was declare her deceased.

In a few short minutes, my whole world collapsed. Nearly half a century of very happy marriage was gone. The head nurse took me to a cold, drab room to wait until I could see her. She was just as beautiful then as when I had first seen her. I called our children. After they heard the news, our dear friend Rene von Richthofen and his wife, Jane Manus, came by the emergency room around 1:30 a.m. to console me. When it was time for me to leave, the police took me home. After all our illnesses, we had always come home together from the hospitals. Now Carolyn was nowhere, and it started to dawn on me that she was gone forever. Now I started to walk on my last lonely road until I would join her.

By the morning, three of our sons were there to support me and to help make arrangements for all we needed to do. Our daughters-in-law in Louisville also started to make arrangements there. Our friends in Palm Beach came by to show their support and brought flowers and trays of good food.

Our next task was to arrange to transport Carolyn's remains to Kentucky for burial. We found out not every flight took caskets. Some airlines were not open for casket arrangement on Sundays, and Monday was Martin Luther King, Jr., Day. Only a large private plane could take the casket, but the expense for it was prohibitive. Finally, all the arrangements had been coordinated, and we flew back to Kentucky with Carolyn's remains by way of Delaware.

Chris's wife, Carolyn, managed much of what needed to be done back at home with much attention given to the smallest details, especially regarding the funeral mass. The visitation was overwhelming. Nearly 400 people came. My nephew, Dr. Stefan Makk, flew in from Austria. Distant and many local friends came from Texas, Palm Beach, and Alabama. I was in a daze through all of this. At the funeral home, I stood with my grandchildren before the casket, saying goodbye to Carolyn. As we held one another, we all just wept. The funeral mass was at St. Martin's Church with a full choir and musicians from the Louisville Orchestra. Then Carolyn was laid to rest in a 160-year-old cemetery on a beautiful hillside.

After the funeral had ended, and the friends and family had left, it started to sink in again that my beloved Carolyn would not be with me ever again in this life.

My family and particularly my grandchildren's love and care were beginning to give me new purpose in life. I had been blessed with the most wonderful family. In this crisis just like every other, they were there and always would be there—loving, caring, and supporting me.

All of Carolyn's doctors, my children who were doctors, and many of our doctor friends were unanimous in their opinion that the Lord took her to spare her the pain and suffering of cancer patients in the terminal stage of that horrible disease. Unfortunately, it was not possible to arrange for a postmortem exam in Florida because of the holiday weekend. Most every doctor felt she passed away due to a pulmonary embolus which is frequently the case in cancer patients.

The loving care of each member of our family and friends was deeply touching. I slowly realized I couldn't keep feeling sorry for myself. My family was replacing my beloved Carolyn's love and affection as much as possible.

Chapter XXVII

The Family

I am blessed with the most wonderful family. For nearly a half a century, I was married to my beautiful and caring wife, Carolyn, until she passed away in January of 2010. She was a talented perfectionist as she ran our household and pursued her interests in interior design and cooking. One could not have asked for a more loving wife, mother, and grandmother.

Our oldest son, Laszlo John Kirtley Makk, is now a dedicated and loving husband and father of three. Recognized as one of the top gastrointestinal specialists in the Louisville area, he shows exemplary dedication to his field and those he serves, impressing his colleagues and patients alike. After he graduated cum laude with a bachelor's degree from Tulane University, he obtained his degree in medicine from the University of Louisville's Medical School.

After the younger Laszlo married and moved out of our home, we found in his room all of the checks that I had paid him when he was a teenager for farm work—checks that had not been cashed. When I reminded him of this, he answered, "Dad, you didn't get paid for your work on the farm, so I felt I shouldn't get paid either." That was coming from a teenager!

He is a wonderful caretaker. Without his loving devotion and medical expertise every time I fell critically ill, I would not be alive today. Once I was discussing pension plans with colleagues when one of them told me that I didn't need a pension plan at all, because I had Laz.

His wife, Ginger, is a registered nurse, a devoted wife and mother, and a talented homemaker. Their oldest child, Laci (seven years old), is a wonderful, loving boy who is very anxious to help anyone with any type of chore. His younger brother, Lucas (six years old), is just like him. He is also very curious and loves to explore everything. Just like his father and uncles were, they are avid fishermen of crawdads in the creek in our front yard. It is music to my heart when the children run into our kitchen asking for pieces of bacon to use as bait for fishing crawdads before running off to the creek. Their little sister, Sophia (four years old), is a beautiful girl and a bit of a tomboy. Her grandmother, Carolyn, designed homemade bracelets and necklaces for her, and she wears them proudly.

Our second son, Stephen Paul Makk, is also a dedicated and loving husband and father of three. A graduate of Tulane University with both a bachelor's in economics and a bachelor's in biology and a graduate of the University of Louisville's Medical School, Steve is a nationally acclaimed orthopedic surgeon.

Because he realized the growing importance that doctors obtain management knowledge, he enrolled in the MBA program at the Kellogg School of Business at Northwestern University. For two years, he flew there every second weekend and managed his on-call duties back at home in between. He certainly made personal sacrifices in order to obtain this additional degree.

When Steve was little, we took the family to Kentucky Dam Village for a vacation. There, we found a little petting zoo. As we approached the pigpen, the pig jumped out, scaring Steve and causing him to climb up a tree. At this time, he wanted to be a farmer. I asked him, "What kind of a farmer would you be if you are afraid of pigs?"

He said, "A fruit farmer." He has been good to his word. When his busy schedule allows, he loves to work in the vineyard and orchard at our farm.

Steve and his wife Melissa have three children. His oldest child, Olivia (16 years old), is a beautiful young lady who excels in academics and works hard at everything she does. She took a babysitting course and saves the money she earns. An "A" student, Olivia was recently elected to be president of the Beta Club at her high school. This organization teaches students how to get involved in the community by helping those who are less fortunate. Olivia has always had a generous heart. When she was just two years old, she visited her grandmother to cheer her up after a chemotherapy treatment. She had even saved her own cinnamon roll to give to her grandma. At first, Olivia didn't recognize her grandmother, because Carolyn had lost her hair from the chemo which scared her;

however, Olivia's good-heartedness overcame her fear, and she was able to give her the treat and visit with Carolyn.

Olivia has two younger brothers, Davis (12 years old) and Hunter (10 years old), who are both wonderful young men. They are good students and excel in several sports, including swimming and soccer. "Study hard, play hard" would be a good motto for them. They are strong boys who can do a man's job at our farm. They are also caring in their ways. During his grandmother's funeral, Hunter just hugged me and sobbed. Several years ago, Davis and Hunter saw that when their neighbor who had Alzheimer's took the garbage out to the end of his drive, he could not find his way back to his house. Without being asked, Davis and Hunter began to take out the neighbor's garbage can from then on.

Christopher Lee Makk is our third son. He is the warmest loving son. A skilled horseman, he was captain of Tulane's polo team. He graduated from Tulane University with a bachelor's in economics and went into the field of logistics and supply chain management with a focus in the food industry. Given his love for food, he is an excellent chef and gardener. His pet project at our farm is raising truffles. This is not all. He is the kindest and most caring person I have ever met. He has cooked for us for years, especially during these last years with his mother in declining health.

He is considerate in many others ways, too. Years ago, our Jack Russell caught a squirrel and started tearing it apart in our backyard just outside the kitchen window. When his mother (Carolyn) told him to chase the dog away, Chris said "I don't want to disturb his meal". Now that's considerate.

Chris's wife, Carolyn, is a wonderful daughter-in-law. She has an MBA from the University of Louisville, and has worked in marketing communications while raising a family. Like Chris, she has always been there to help us in our times of need, and she continues to be there for me today.

Their oldest son, Marshall (15 years old), has a brilliant mind, and yet he is very humble. He is an "A" honors student, and he is like a walking encyclopedia. He just recently received an invitation to audition for the *Jeopardy's* teen competition. He also has a gift and a passion for theater, and he has been involved with the nationally known Walden Theatre for young people for many years.

Their youngest, Matthew (13 years old), is a kind and caring grandson who is also an "A" student and quite gifted. He has a talent and passion for music. He is a gifted musician who is mastering piano and guitar.

However, he enjoys any instrument he picks up. He plans to audition for the Youth Performing Arts School for high school. He is a hard worker and very industrious, cutting grass and shoveling snow from driveways to earn spending money to buy musical instruments.

Our youngest son, Andrew Francis Makk, followed in the family tradition and obtained a BSM in finance from Tulane University. He later earned an MBA from the Fuqua School of Business at Duke University. He was a wonderful, playful baby and a lot of fun as he was growing up. Every evening when I came home, he used to play and play what we called our standard games. When I became chief pathologist at St. Anthony's and Methodist Hospital, I had a lot of evening meetings. One evening, I asked Andrew why we didn't play our games anymore. He said, "Dad, I wait for you every night, but you always have a meeting." This struck a chord in my heart, and I told Carolyn come July 1st, I would give up one of my two positions.

When Andrew was an eighth grader, he had a horrible accident in school and broke his right femur at the top of his leg. He was in traction for six weeks and then in a total body cast all summer. His friends visited for a while, but they didn't know what to say or do and stopped coming. He lay stiff like a mummy on the playroom couch but never complained. I bought him a computer to keep him busy. He acquired a lot of computer skills on his own and even designed games on it —all while lying flat in a body cast. He recovered and even played football after he returned to school.

Andrew has always been there for us in his special way as the youngest child. He has come home to Louisville and stayed with us during our different medical crises. Always focused on career as a hard worker, Andrew's first job was with Enron, where he became the company's youngest vice president at the age of 29. Enron fell apart and he suddenly lost his job, Andrew was not bitter about it. He learned much from the experience, and he is now a principal at Energy Capital Partners. He loves his job, but he is also happily married and a proud new father of a beautiful baby girl.

Andrew's wife, Catherine, is a lovely and smart person. She most recently worked as executive marketing director for *Gourmet* magazine. She was wonderful to my wife during her struggle with cancer. Their little girl was born just a week after Carolyn's funeral. They named her Hannah Carolyn Makk. She is adorable and holds a special place in my heart. I am sure Carolyn would have been delighted just like me.

Every day, Carolyn and I realized how blessed we were with our children and grandchildren in our lives.

The family would not be complete without including John and Faye Bender. They took me into their home when I arrived in Albany, New York, as an immigrant. Pretty soon, I had the privilege and good fortune of being treated like a family member, which continued for decades to come.

John Bender had offered to put me through law school at Yale, a Bender family tradition, if I committed to working for them at their book publishing company. I turned this most generous and comforting offer down, because I was determined to become a doctor.

The Benders were caring and supportive throughout my medical school years. They treated my wife, Carolyn, like a true daughter-in-law, and our children like their own grandchildren. They invited us to take part in their family vacations, and they even treated us all to a cruise one year. They moved to Palm Beach, and we spent time with them there.

Over the years, we have been able to reciprocate some of this family love and support. On the 40th anniversary of my coming to this great country, we honored the Benders at an elegant party at the Everglades Club in Palm Beach.

When John Bender passed away, I had the honor of being one of the speakers at his funeral service. As time passed on, his wife, Faye, became frailer. It was a special pleasure for us to take her out for dinner or visit her and keep her company. The last time we visited her, she was in no shape to go anywhere. We just had a quiet meal with her and knew she was ready to pass on. It was not long before she, too, was gone. Her kindness, love, and generosity will always remain with us. May the Lord keep Faye and John Bender in His graces.

Steve Tubbs is not a blood relative, but he is just like family. He has managed our farm (Six Acorn Farm) for nearly a quarter of a century. The property and livestock are always taken care of even during severe weather and cold winters. For five years, Steve drove Carolyn to her chemotherapy sessions in Cincinnati, Ohio. I would accompany them and was touched by his gentle and supportive ways, always trying to make Carolyn smile and ease the strain of her burden. His lovely wife, Edith, always had a gift for Carolyn. Steve's helper, Jimmy Combs, is an honest and wonderful person as well. We can be away from the farm for long periods of time and know that he is there keeping an eye on things and helping out.

My wonderful family and friends lessen my grief. They have always come through loving and helping in the past when we had a health scare or some other type of crisis. The remainder of my life won't be a lonely path to travel, as they will always be there with their love and care.

God bless my family.
God bless my friends.
And God bless America.

Our struggle goes on.
Our hope never dims.
God lights the road.
We are traveling.

Appendix I: Awards

Dr. Laszlo Makk is an award-winning physician equally recognized for his contributions by his peers, his community, and his country. He has received many awards including the following:

St. Anthony's Medical Center "Distinguished Service Award"

In 1983, St. Anthony's Medical Center (Louisville, KY) awarded Dr. Makk with the Distinguished Service Award. Dr. Makk is one of only four physicians to receive this award in the hospital's history. His citation stated:

> *"In recognition of his local, national, and international leadership and tradition of excellence in the field of clinical and anatomic pathology, to a man who personifies the qualities and highest ideals as a physician, researcher, writer, friend, and American, we especially recognize his service and loyalty to the hospital and to the medical community. We are grateful and honored to know a person of his distinction who cheerfully shares himself and his knowledge with all, both professional and laymen alike."*

Daughters of the American Revolution "Americanism Award"

In 1996, Dr. Makk received the DAR Americanism Award. This award is given to an adult naturalized American citizen who has shown the attributes of trustworthiness, service, leadership, and patriotism to a great extent.

When asked to produce a letter of recommendation as part of the nomination process for this award, Dr. William Vonder Haar, chairman of the department of family practice at the University of Louisville, had this to say about Laszlo at the time:

> *I am pleased to be a part of any effort to recognize the patriotism, dedication to the United States of America and what it stands for, trustworthiness, and compassion of Dr. Laszlo Makk ... He has constantly reminded those of us who have taken our American freedom for granted that we need to revere and honor it.*

Prominent Louisville attorney William Stodghill recommended Laszlo for the award as well, stating in his letter:

> *Laszlo and I first met some twenty years ago as members of the Louisville Downtown Rotary Club. Shortly after Laszlo joined Rotary, he presented a talk vividly describing his exodus from Hungary at the time of the Russian takeover and indelibly impressed upon all of us not only the courage it takes to leave one's country but also the unique and precious freedoms that we enjoy as Americans.*

American Hungarian Foundation "Abraham Lincoln Award"

In 1999, Dr. Makk received the Abraham Lincoln Award from the American Hungarian Foundation, which was presented by AHF's President August Molnar.

[Dr. Makk has been the recipient of many other civic and professional awards, and is the author of more than 50 medical publications. In addition, he is a happy farmer.]